Newnes
Mathematics
Pocket Book
for engineers

J O Bird
BSc(Hons), AFIMA, TEng(CEI), MIElec IE

D1427719

Newnes Technical Books

Newnes Technical Books
is an imprint of the Butterworth Group
which has principal offices in
London, Boston, Durban, Singapore, Sydney, Toronto, Wellington

First published 1983

© Butterworth & Co (Publishers) Ltd 1983

British Library Cataloguing in Publication Data

Bird, J. O.
  Newnes mathematics pocket-book for engineers.
  1. Engineering mathematics
  I. Title
510′.2462   TA330

  ISBN 0-408-01330-3

Photoset by Mid-County Press, London SW15
Printed in England by The Thetford Press Ltd, Thetford, Norfolk

Preface

This Mathematics pocket book provides students, technicians, engineers and scientists with a readily available reference to the essential mathematics formulae, definitions and general information needed during their studies and/or work situation.

The book assumes little previous knowedge, is suitable for a wide range of courses, and will be particularly useful for students studying for Technician certificates and diplomas, and for C.S.E. and 'O' and 'A' levels.

The author would like to express his appreciation for the friendly cooperation and helpful advice given to him by the publishers and to Mr Tony May for his assistance in editing the manuscript. Thanks are also due to Mrs Elaine Woolley for the excellent typing of the manuscript.

Finally, the author would like to add a word of thanks to his wife Elizabeth for her patience, help and encouragement during the preparation of this book.

J O Bird
Highbury College of Technology
Portsmouth

Contents

1 Arithmetic operations

1 **Arithmetic** is the science of numbers — the art of counting, computing or calculating.

HCF

2 When two or more numbers are multiplied together, the individual numbers are called factors. Thus a factor is a number which divides into another number exactly. The **highest common factor (HCF)** is the largest number which divides into two or more numbers exactly.

For example, to find the HCF of 12, 30 and 42, firstly each number is expressed in terms of its lowest factors. This is achieved by repeatedly dividing by the prime numbers 2, 3, 5, 7, 11, 13, ..., (where possible) in turn. Thus

$$12 = 2 \times 2 \times 3$$
$$30 = 2 \qquad \times 3 \times 5$$
$$42 = 2 \qquad \times 3 \qquad \times 7$$

The factors which are common to each of the numbers are 2 in column 1 and 3 in column 3, shown by the broken lines. Hence the HCF is 2×3, i.e. 6. That is, 6 is the largest number which will divide into all of the given numbers.

LCM

3 A multiple is a number which contains another number an exact number of times. The smallest number which is exactly divisible by each of two or more numbers is called the **lowest common multiple, (LCM)**.

For example, to find LCM of 12, 42 and 90, each number is expressed in terms of its lowest factors, and then selecting the largest group of any of the factors present. Thus,

$$12 = 2 \times 2 \times 3$$
$$42 = 2 \qquad \times 3 \qquad \times 7$$
$$90 = 2 \qquad \times 3 \times 3 \times 5$$

The largest group of any of the factors present are shown by the broken lines and are 2×2 in 12, 3×3 in 90, 5 in 90 and 7 in 42. Hence the LCM is $2 \times 2 \times 3 \times 3 \times 5 \times 7 = 1260$, and is the smallest number which 12, 42 and 90 will all divide into exactly.

Order of Precedence

4   In arithmetic operations, the order in which operations are performed are:

> (i) to determine the values of operations contained in brackets,
> (ii) multiplication and division, (the word "of" also means multiply), and
> (iii) addition and subtraction.

This order of precedence can be remembered by the word **BODMAS**, standing for **B**rackets, **O**f, **D**ivision, **M**ultiplication, **A**ddition and **S**ubtraction, taken in that order.

Thus, for example,

$$13 - 2 \times 3 + 14 \div (2 + 5)$$

is evaluated as:

$13 - 2 \times 3 + 14 \div 7$	(**B**rackets)
$= 13 - 2 \times 3 + 2$	(**D**ivision)
$= 13 - 6 + 2$	(**M**ultiplication)
$= 15 - 6$	(**A**ddition)
$= \mathbf{9}$	(**S**ubtraction)

Fractions

5   When 2 is divided by 3, it may be written as $\frac{2}{3}$ or 2/3. $\frac{2}{3}$ is called a fraction. The number above the line, i.e. 2, is called the **numerator** and the number below the line, i.e. 3, is called the **denominator**. When the value of the numerator is less than the value of the denominator, the fraction is called a proper fraction; thus $\frac{2}{3}$ is a **proper fraction**. When the value of the numerator is greater than the denominator, the fraction is called an **improper fraction**. Thus $\frac{7}{3}$ is an improper fraction and can also be expressed as a **mixed number**, that is, an integer and a proper fraction. Thus the improper fraction $\frac{7}{3}$ is equal to the mixed number $2\frac{1}{3}$.

When a fraction is simplified by dividing the numerator and denominator by the same number, the process is called **cancelling**. Cancelling by 0 is not permissable.

Continued fractions

6   Any fraction may be expressed in the form shown below for the fraction $\frac{26}{55}$.

$$\frac{26}{55} = \frac{1}{\frac{55}{26}} = \frac{1}{2+\frac{3}{26}} = \frac{1}{2+\frac{1}{\frac{26}{3}}} = \frac{1}{2+\frac{1}{8+\frac{2}{3}}} = \frac{1}{2+\frac{1}{8+\frac{1}{\frac{3}{2}}}} = \frac{1}{2+\frac{1}{8+\frac{1}{1+\frac{1}{2}}}}$$

The latter form can be expressed generally as

$$\cfrac{1}{A+\cfrac{\alpha}{B+\cfrac{\beta}{C+\cfrac{\gamma}{D+\delta}}}}$$

Comparison shows that A, B, C and D are 2, 8, 1 and 2 respectively.

A fraction written in the general form is called a **continued fraction** and the integers A, B, C and D are called the **quotients** of the continued fraction. The quotients may be used to obtain closer and closer approximations to the original fraction, these approximations being called **convergents**.

A tabular method may be used to determine the convergents of a fraction:

		1	2	3	4	5
a			2	8	1	2
b $\begin{cases} \frac{bp}{bq} \end{cases}$		$\frac{0}{1}$	$\frac{1}{2}$	$\frac{8}{17}$	$\frac{9}{19}$	$\frac{26}{55}$

(i) The quotients 2, 8, 1 and 2 are written in cells a2, a3, a4 and a5 with cell a1 being left empty.

(ii) the fraction $\frac{0}{1}$ is always written in cell b1.

(iii) The reciprocal of the quotient in cell a2 is always written in cell b2 i.e. $\frac{1}{2}$ in this case.

(iv) The fraction in cell b3 is given by $\dfrac{(a3 \times b2p) + b1p}{(a3 \times b2q) + b1q}$,

i.e. $\dfrac{(8 \times 1) + 0}{(8 \times 2) + 1} = \dfrac{8}{17}$

(v) The fraction in cell b4 is given by $\dfrac{(a4 \times b3p) + b2p}{(a4 \times b3q) + b2q}$,

$= \dfrac{(1 \times 8) + 1}{(1 \times 17) + 2} = \dfrac{9}{19}$, and so on.

Hence the convergents of $\dfrac{26}{55}$ are $\dfrac{1}{2}$, $\dfrac{8}{17}$, $\dfrac{9}{19}$ and $\dfrac{26}{55}$,

each value approximating closer and closer to $\dfrac{26}{55}$.

These approximations to fractions are used to obtain practical ratios for gearwheels or for a dividing head (used to give a required angular displacement).

Ratio

7   The ratio of one quantity to another is a fraction, and is the number of times one quantity is contained in another quantity **of the same kind**. Thus expressing 25p as a ratio of £3.75 is

$\dfrac{25}{375}$, i.e. $\dfrac{1}{15}$

Proportion

8   If one quantity is **directly proportional** to another, then as one quantity doubles, the other quantity also doubles. When a quantity is **inversely proportional** to another, then as one quantity doubles, the other quantity is halved. (See section 3, para. 5, page 25).

If it is required to divide 126 in the ratio of 5 to 13, then firstly the parts are added together, i.e., $5 + 13 = 18$. Hence 18 parts corresponds to 126.

thus   1 part corresponds to $\dfrac{126}{18} = 7$

5 parts corresponds to $5 \times 7 = 35$

and 13 parts corresponds to $13 \times 7 = 91$.

Thus 126 divided in the ratio of 5:13 is 35:91.

Decimals

9   The decimal system of numbers is based on the **digits** 0 to 9. A number such as 53.17 is called a **decimal fraction**, a **decimal point** separating the integer part, i.e. 53, from the fractional part, i.e. 0.17.

A number which can be expressed exactly as a decimal fraction is called a **terminating decimal** and those which cannot be expressed exactly as a decimal fraction are called **non-terminating decimals**.

Thus, $\frac{3}{2} = 1.5$ is a terminating decimal, but $\frac{4}{3} = 1.33333\ldots$ is a non-terminating decimal.

$1.33333\ldots$ can be written as $1.\dot{3}$, called 'one point-three recurring'.

The answer to a non-terminating decimal may be expressed in two ways, depending on the accuracy required:

(i) correct to a number of significant figures, that is, figures which signify something, and
(ii) correct to a number of decimal places, that is, the number of figures after the decimal point.

The last digit in the answer is unaltered if the next digit on the right is in the group of numbers 0, 1, 2, 3 or 4, but is increased by 1 if the next digit on the right is in the group of numbers 5, 6, 7, 8 or 9. Thus the non-terminating decimal $7.6183\ldots$ becomes 7.62, correct to 3 significant figures, since the next digit on the right is 8, which is in the group of numbers 5, 6, 7, 8 or 9. Also $7.6183\ldots$ becomes 7.618, correct to 3 decimal places, since the next digit on the right is 3, which is in the group of numbers 0, 1, 2, 3 or 4.

Percentages

10   Percentages are fractions having the number 100 as their denominator and are used to give a common standard.

For example, 25 per cent means $\frac{25}{100}$, i.e. $\frac{1}{4}$, and is written as 25%.

Thus $12\frac{1}{2}\%$ of £378 means $\frac{12\frac{1}{2}}{100} \times £378 = \frac{1}{8} \times £378 = \mathbf{£47.25}$.

Expressing 134 mm as a percentage of 2.5 m is $\frac{134}{2500} \times 100\%$
$$= \mathbf{5.36\%}.$$

Bases, indices and powers

11 The lowest factors of 2 000 are $2 \times 2 \times 2 \times 2 \times 5 \times 5 \times 5$. These factors are written as $2^4 \times 5^3$, where 2 and 5 are called **bases** and the numbers 4 and 3 are called **indices**.

When an index is an integer it is called a **power**.

Thus, 2^4 is called 'two to the power of four', and has a base of 2 and an index of 4. Similarly, 5^3 is called 'five to a power of 3' and has a base of 5 and an index of 3. Special names may be used when the indices are 2 and 3, these being called 'squared' and 'cubed' respectively. Thus 7^2 is called 'seven squared' and 9^3 is called 'nine cubed'. When no index is shown, the power is 1, i.e. 2^1 means 2.

Reciprocals

12 The reciprocal of a number is when the index is -1 and its value is given by 1 divided by the base. Thus the reciprocal of 2 is 2^{-1} and its value is $\frac{1}{2}$ or 0.5. Similarly, the reciprocal of 5 is 5^{-1} which means $\frac{1}{5}$ or 0.2.

Square roots

13 The square root of a number is when the index is $\frac{1}{2}$, and the square root of 2 is written as $2^{1/2}$ or $\sqrt{2}$. The value of a square root is the value of the base which when multiplied by itself gives the number. Since $3 \times 3 = 9$, then $\sqrt{9} = 3$. However, $(-3) \times (-3) = 9$, so $\sqrt{9} = -3$. There are always two answers when finding the square root of a number and this is shown by putting both a + and a − sign in front of the answer to a square root problem. Thus $\sqrt{9} = \pm 3$ and $4^{1/2} = \sqrt{4} = \pm 2$, and so on.

Laws of indices

14 When simplifying calculations involving indices, certain basic rules or laws can be applied, called the laws of indices. These are given below.

(i) When multiplying two or more numbers having the same base, the indices are added. Thus

$$3^2 \times 3^4 = 3^{2+4} = 3^6$$

(ii) When a number is divided by a number having the same base, the indices are subtracted. Thus

$$\frac{3^5}{3^2} = 3^{5-2} = 3^3.$$

(iii) When a number which is raised to a power is raised to a further power, the indices are multiplied. Thus

$$(3^5)^2 = 3^{5 \times 2} = 3^{10}.$$

(iv) When a number has an index of 0, its value is 1. Thus $3^0 = 1$.

(v) A number raised to a negative power is the reciprocal of that number raised to a positive power. Thus

$$3^{-4} = \frac{1}{3^4}.$$

Similarly, $\frac{1}{2^{-3}} = 2^3$.

(vi) When a number is raised to a fractional power the denominator of the fraction is the root of the number and the numerator is the power. Thus, $8^{2/3} = \sqrt[3]{8^2} = (2)^2 = 4$ and $25^{1/2} = \sqrt{25^1} = \pm 5$.

Standard form

15   A number written with one digit to the left of the decimal point and multiplied by 10 raised to some power is said to be wrritten in standard form. Thus:

5 837 is written as 5.837×10^3 in standard form, and 0.0415 is written as 4.15×10^{-2} in standard form.

When a number is written in standard form, the first factor is called the **mantissa** and the second factor is called the **exponent**. Thus the number 5.8×10^3 has a mantissa of 5.8 and an exponent of 10^3.

Errors

16   (i) In all problems in which the measurement of distance, time, mass or other quantities occurs, an exact answer cannot be given; only an answer which is correct to a

stated degree of accuracy can be given. To take account of this an **error due to measurement** is said to exist.
(ii) To take account of measurement errors it is usual to limit answers so that the result given is **not more than one significant figure greater than the least accurate number given in the data.**
(iii) **Rounding-off errors** can exist with decimal fractions. For example, to state that $\pi = 3.142$ is not strictly correct, but '$\pi = 3.142$ correct to 4 significant figures' is a true statement. (Actually, $\pi = 3.14159265...$)
(iv) It is possible, through an incorrect procedure, to obtain the wrong answer to a calculation. This type of error is known as a **blunder**.
(v) An **order of magnitude error** is said to exist if incorrect positioning of the decimal point occurs after a calculation has been completed.

Logarithms

17 (i) A logarithm of a number is the power to which a base has to be raised to be equal to the number.
Thus,

if $y = a^x$ then $x = \log_a y$.

(ii) Logarithms having a base of 10 are called **common logarithms** and the common logarithm of x is written as lg x.
(iii) Logarithms having a base of e are called **hyperbolic**, **Naperian** or **natural logarithms** and the Naperian logarithm of x is written as $\log_e x$, or more commonly, ln x.
(iv) The change of base rule for logarithms states,
$\log_a y = \dfrac{\log_b y}{\log_b a}$, from which, $\ln y = \dfrac{\lg y}{\lg e} = \dfrac{\lg y}{0.4343} = 2.3026 \lg y.$
(v) Laws of logarithms:

(a) $\log (A \times b) = \log A + \log B$

(b) $\log\left(\dfrac{A}{B}\right) = \log A - \log B$

(c) $\log A^n = n \log A$

8

2 Numbering systems

Conversion of denary numbers to binary numbers and vice-versa

1   (i) The system of numbers in everyday use is the denary or decimal system of numbers, using the digits 0 to 9. It has ten different digits (0, 1, 2, 3, 4, 5, 6, 7, 8 and 9), and is said to have a radix or base of 10.

(ii) the binary system of numbers has a radix of 2 and uses only the digits 0 to 1.

2   (i) The denary number 234.5 is equivalent to

$$2 \times 10^2 + 3 \times 10^1 + 4 \times 10^0 + 5 \times 10^{-1}$$

i.e., is the sum of terms comprising: (a digit) multiplied by (the base raised to some power).

(ii) In the binary system of numbers, the base is 2, so 1 101.1 is equivalent to:

$$1 \times 2^3 + 1 \times 2^2 + 0 \times 2^1 + 1 \times 2^0 + 1 \times 2^{-1}$$

Thus the denary number equivalent to the binary number 1 101.1 is

$$8 + 4 + 0 + 1 + \frac{1}{2}, \text{ that is } 13.5,$$

i.e. $1\,101.1_2 = 13.5_{10}$, the suffixes 2 and 10 denoting binary and denary systems of numbers respectively.

3   An integer denary number can be converted to a corresponding binary number by repeatedly dividing by 2 and noting the remainder at each stage, as shown below for 39_{10}.

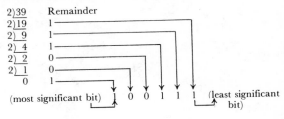

(most significant bit) 1 0 0 1 1 1 (least significant bit)

The result is obtained by writing the top digit of the remainder as the least significant bit, (a bit is a binary digit and the least significant bit is the one on the right). The bottom bit of the remainder is the most significant bit, i.e. the bit on the left.

Thus $39_{10} = 100\,111_2$.

4   The fractional part of a denary number can be converted to a binary number by repeatedly multiplying by 2, as shown below for the fraction 0.625.

$$
\begin{array}{lll}
0.625 \times 2 = & 1. & 250 \\
0.250 \times 2 = & 0. & 500 \\
0.500 \times 2 = & 1. & 000
\end{array}
$$

(most significant bit) .1 0 1 (least significant bit)

For fractions, the most significant bit of the result is the top bit obtained from the integer part of multiplication by 2. The least significant bit of the result is the bottom bit obtained from the integer part of multiplication by 2. Thus $0.625_{10} = 0.101_2$.

5   For denary integers containing several digits, repeatedly dividing by 2 can be a lengthy process. In this case, it is easier usually to convert a denary number to a binary number via the octal system of numbers. This system has a radix of 8, using the digits 0, 1, 2, 3, 4, 5, 6 and 7. The denary number equivalent to the octal number $4\,317_8$ is

$$4 \times 8^3 + 3 \times 8^2 + 1 \times 8^1 + 7 \times 8^0$$
i.e. $4 \times 512 + 3 \times 64 + 1 \times 8 + 7 \times 1$ or $2\,255_{10}$.

6   An integer denary number can be converted to a corresponding octal number by repeatedly dividing by 8 and noting the remainder at each stage, as shown below for 493_{10}.

$$
\begin{array}{ll}
8)\underline{493} & \text{Remainder} \\
8)\underline{\ 61} & 5 \\
8)\underline{\ \ 7} & 5 \\
\ \ \ \ 0 & 7
\end{array}
$$
                    7 5 5

Thus, $493_{10} = 755_8$.

7   The fractional part of a denary number can be converted to an octal number by repeatedly multiplying by 8, as shown below for the fraction 0.4375_{10}.

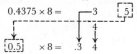

$$
\begin{array}{ll}
0.4375 \times 8 = & 3 \qquad 5 \\
& 4 \\
0.5 \quad \times 8 = & .3 \quad 4
\end{array}
$$

For fractions, the most significant bit is the top integer obtained by multiplication of the denary fraction by 8. Thus

$$0.4375_{10} = 0.34_8$$

Table 2.1

Octal digit	Natural binary number
0	000
1	001
2	010
3	011
4	100
5	101
6	110
7	111

8   The natural binary code for digits 0 to 7 is shown in *Table 2.1*, and an octal number can be converted to a binary number by writing down the three bits corresponding to the octal digit.

Thus, $437_8 = 100\ 011\ 111_2$
and $26.35_8 = 010\ 110.011\ 101_2$

The '0' on the extreme left does not signify anything, thus

$$26.35_8 = 10\ 110.011\ 101_2$$

To convert a denary number to a binary number via octal, the denary number is first converted to an octal number, as shown in paras 6 and 7, and then the corresponding binary number is written down, as shown above.

Addition and subtraction of binary numbers

9   Binary addition of two bits is achieved according to the following rules:

	Sum	Carry
$0 + 0 =$	0	0
$0 + 1 =$	1	0
$1 + 0 =$	1	0
$1 + 1 =$	0	1

When adding binary numbers A and B, A is called the augend and B is called the addend. For example, adding 1 and 1 produces a sum of 0 and a carry of 1, and may be laid out as:

Augend	1
Addend	1
Sum	0
Carry 1	

10 Binary addition of three bits, (augend + addend + carry) is achieved according to the following rules:

		Sum	Carry
$0+0+0$	=	0	0
$0+0+1$	=	1	0
$0+1+0$	=	1	0
$0+1+1$	=	0	1
$1+0+0$	=	1	0
$1+0+1$	=	0	1
$1+1+0$	=	0	1
$1+1+1$	=	1	1

Using the rules of two bit and three bit addition, two binary numbers say 1011 and 1110, may be added as shown:

Column	5	4	3	2	1
Augend	0	1	0	1	1
Addend	0	1	1	1	0
Sum	1	1	0	0	1
Carry 0	1	1	1	0	

Column 1: Adding augend and addend gives $1+0=$ sum 1, carry 0 to column 2.
Column 2: Adding augend, addend and carry gives: $1+1+0$ = sum 0, carry 1 to column 3.
Column 3: Adding augend, addend and carry gives: $0+1+1$ = sum 0, carry 1 to column 4, and so on.

11 A negative binary number can be stored in a calculator or computer by using a sign bit to denote the negative quantity. Binary numbers having a sign bit are called **signed-magnitude** binary numbers. An additional bit is allocated to the left of the binary number to indicate the sign and by convention, a sign bit of 0 is used to denote a positive number and a 1 to denote a negative number. Thus, in a signed-bit system, (0) 1001_2 represents the denary number $+9$, but (1) 1001_2 represents the denary number -9, the sign bit being shown in brackets.

Signed-magnitude binary numbers have two parts:

(i) the sign bit (shown in brackets in this section for clarity), giving the sign of the number, and
(ii) a binary number following the sign bit giving the size or **modulus** of the binary number.

One of the disadvantages of a signed-bit system is that it reduces the maximum number capable of being stored in a given sized register by approximately a half. For example, a 4-bit register can store the positive numbers 0 to 15 if no sign bit is used, and the numbers -7 to $+7$ when a sign bit is used.

12   In many calculators, microprocessors and computers, the process of subtracting one binary number from another is performed by a process called complementary addition. This process enables the same logic circuitry to be used for both addition and subtraction. There are two methods of complementary addition in widespread use: (i) the one's complement method, shown in para 13, and (ii) the two's complement method, shown in para 14.

The one's complement method

13   When subtracting binary number B from binary number A, i.e. $A - B = C$, A is called the **minuend**, B is called the **subtrahend** and C is called the **difference**, i.e. **minuend − subtrahend = difference**. The procedure for subtracting one binary number from another using the one's complement method is as follows:

(i) Express both minuend and subtrahend so that they each have the same number of bits.
(ii) Determine the one's complement of the subtrahend. This is achieved by writing 1 for 0 and 0 for 1 for each bit in the subtrahend. Thus the one's complement of 101 101 is 010010.
(iii) Add the minuend to the one's complement of the subtrahend to obtain a sum.
(iv) Depending on the number of bits in the sum, the result of complementary addition, using the one's complement method is obtained thus:

(a) If the sum has the same number of bits as the minuend and subtrahend of (i), the difference between the minuend and subtrahend is negative, and the value is the one's complement of the sum.
(b) If one extra bit is generated in the sum by complementary addition in (iii), the difference between the minuend and

subtrahend is positive and its value is given by **end-around carry**, that is, by taking the bit on the extreme left of the sum and adding it to the bit on the extreme right. Thus, end-around carry operating on $1\,001\,011$ gives $1\,100$ as shown:

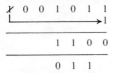

Thus, for example, to determine $1011_2 - 110_2$ by the one's complement method the above procedure is followed for a minuend of 1011 and a subtrahend of 110.

(i) $1011 - 110 = 1011 - 0110$

(ii) The one's complement of the subtrahend is 1001.

(iii) Adding the minuend and the one's complement of the subtrahend gives:

$$\begin{array}{r} 1\,011 \\ 1\,001 \\ \hline \end{array}$$

Sum $\quad 10\,100$

(iv) An extra bit is generated in the sum so the procedure given in (iv)(b) above is adopted. The difference is positive and its value is obtained by 'end-around carry', i.e.,

$$\begin{array}{c} \cancel{1}0\,100 \\ \llcorner\!\!\rightarrow 1 \end{array}$$

Sum $\quad 101$ Thus $\mathbf{1\,011_2 - 110_2 = 101_2}$.

The two's complement method

14   Assuming the subtraction is of the form:

minuend − subtrahend = difference,

then the procedure for determining the difference by the two's complement method is as follows:

(i) Express the minuend and subtrahend so that they each have the same number of bits.

(ii) Add the sign-bit, (0) for a positive number and (1) for a negative number on the extreme left of both the

14

minuend and subtrahend. (*Note*: $10\,111 - 101$ means that both minuend and subtrahend are positive numbers, i.e. $(+10\,111) - (+101)$ and are assigned sign-bits of (0)).

(iii) Determine the two's complement of the subtrahend. This is achieved by writing down the one's complement and adding 1. Thus the two's complement of $101\,011$ is $010\,100 + 1$, i.e. $010\,101$.

(iv) Add the minuend to the two's complement of the subtrahend to obtain a sum.

(v) Depending on whether the sign bit is (0) or (1), the result of complementary addition by the two's complement method is obtained thus:

(a) If the sign bit is (0), the 1 on the left of the sign is disregarded and the difference between the minuend and subtrahend is positive and its value is given by the bits to the right of the sign bit in the sum.

(b) If the sign bit is (1), the difference between the minuend and subtrahend is negative and its value is the two's complement of the sum.

Thus, for example, to determine $1\,101\,011_2 - 110\,100_2$ by the two's complement method the above procedure is followed for a minuend of $1\,101\,011$ and a subtrahend of $110\,100$.

(1) $1\,101\,011 - 110\,100 = 1\,101\,011 - 0\,110\,100$.

(ii) Adding the sign bits gives $(0)1\,101\,011 - (0)0\,110\,100$, the sign bit being (0) since both numbers are positive, i.e.,

$$(+1\,101\,011_2) - (+110\,100_2).$$

(iii) The two's complement of the subtrahend is obtained from the one's complement plus one, i.e. $(1)1\,001\,011 + 1$, that is, $(1)1\,001\,100$.

(iv) Adding the minuend and two's complement of the subtahend gives:

$$(0)1\,101\,011$$
$$(1)1\,001\,100$$

Sum $\quad 1(0)0\,110\,111$

(v) The sign bit is (0), hence (v)(a) of the above procedure applies. The 1 on the left of the sign bit is disregarded, the difference is positive and its value is given by the bits on the right of the sign bit.

Thus **$1\,101\,011_2 - 110\,100_2 = 110\,111_2$**

15   **Binary subtraction** of two numbers may also be achieved by a method of direct subtraction, according to the rules:

		difference	borrow
$0 - 0$	=	0	0
$0 - 1$	=	1	1
$1 - 0$	=	1	0
$1 - 1$	=	0	0

Thus, the method of taking 1011_2 from 1101_2 is as shown, the subtraction for each column being:
minuend $-$ (subtrahend $+$ borrow).

Column	5	4	3	2	1
Minuend		1	1	0	1
Subtrahend		1	0	1	1
Difference		0	0	1	0
Borrow	0	0	1	0	0

Column 1: $1 - (1 + 0) =$ difference 0, borrow 0 (in column 2).
Column 2: $0 - (1 + 0) =$ difference 1, borrow 1 (in column 3).
Column 3: $1 - (0 + 1) =$ difference 0, borrow 0 (in column 4).
Column 4: $1 - (1 + 0) =$ difference 0, borrow 0 (in column 5).

Thus **$1101_2 - 1011_2 = 10_2$**.

In practice, the procedure is much the same as subtraction in denary numbers. To take 1011_2 from 1101_2 is as shown:

Column	4	3	2	1
Minuend	1	0	2	1
Subtrahend	1	0	1	1
Difference	0	0	1	0

Column 1: $1 - 1 =$ difference 0.
Column 2: $0 - 1$, borrow 1 from column 3 leaving the minuend of column 3 as 0. The borrowed 1 becomes 2 when moved to column 2. $2 - 1 =$ difference 1.
Column 3: $0 - 0 =$ difference 0.   Thus, **$1101_2 - 1011_2 = 10_2$**,
Column 4: $1 - 1 =$ difference 0.   as obtained previously.

Multiplication and division of binary numbers

16   Binary multiplication of two bits is according to the following rules:

$$0 \times 0 = 0$$
$$0 \times 1 = 0$$
$$1 \times 0 = 0$$
$$1 \times 1 = 1$$

When multiplying A by B to give C, i.e., $A \times B = C$, A is called the **multiplicand**, B the **multiplier** and C the **product**.

17   The operation of multiplication in binary is similar to that for denary numbers, as shown below for $14_{10} \times 6_{10}$.

```
14₁₀  =              1  1  1  0₂   (multiplicand)
 6₁₀  =              0  1  1  0₂   (multiplier)
                  ─────────────
                 0  0  0  0
              1  1  1  0
           1  1  1  0
        0  0  0  0
                  ─────────────
84₁₀  =  1  0  1  0  1  0  0₂   (product)
```

It can be shown that zero bits in the multiplier do not contribute to the product and are usually ignored. Thus, binary multiplication of $25_{10} \times 13_{10}$ is performed as shown.

```
25₁₀  =           1  1  0  0  1   (multiplicand)
13₁₀  =           1  1  0  1      (multiplier)
               ──────────────
               1  1  0  0  1
            1  1  0  0  1
         1  1  0  0  1
               ──────────────
325₁₀ =  1  0  1  0  0  0  1  0  1   (product)
```

18   Denary multiplication of numbers having decimal fractions is performed by ignoring the decimal point in the first instance and then counting the number of fractional digits in both the multiplicand and multiplier to position the decimal point in the product. Thus:

$$31.\underline{4}_{10} \times 13.\underline{72}_{10} = 430.\underline{808}_{10}$$

A similar procedure is used for binary numbers. Thus

$\left(12\frac{1}{2}\right)_{10} \times \left(9\frac{3}{8}\right)_{10}$ in binary multiplication gives:

$$\left(12\frac{1}{2}\right)_{10} = \qquad\qquad 1 \quad 1 \quad 0 \quad 0 \;.\; \underline{1} \qquad\qquad \text{(multiplicand)}$$

$$\left(9\frac{3}{8}\right)_{10} = \qquad\qquad 1 \quad 0 \quad 0 \;.\; \underline{1} \quad \underline{0} \quad \underline{1} \quad \underline{1} \qquad \text{(multiplier)}$$

```
                          1   1   0   0   1
                      1   1   0   0   1            (partial
                  1   1   0   0   1                 products)
          1   1   0   0   1
```

$$117.187\,5_{10} = \quad 1 \quad 1 \quad 1 \quad 0 \quad 1 \quad 0 \quad 1 \;.\; \underline{0} \quad \underline{0} \quad \underline{1} \quad \underline{1} \quad \text{(product)}$$

(Should it occur, when adding the partial products, then
$1+1+1+1=0$, carry 2; $1+1+1+1+1=1$, carry 2; and so on.)

19   The process of multiplication can be performed by adding the multiplicand to itself as many times as required by the multiplier, thus

$$9_{10} \times 4_{10} = 9_{10} + 9_{10} + 9_{10} + 9_{10} = 36_{10}$$

Similarly, in binary numbers

$$1011_2 \times 101_2 = 1011_2 + 1011_2 + 1011_2 + 1011_2 + 1011_2 \quad (\text{since } 101_2 = 5_{10})$$

Using the principles introduced in paras 9 and 10:

```
          0   0   1   0   1   1
          0   0   1   0   1   1
          0   0   1   0   1   1
          0   0   1   0   1   1
          0   0   1   0   1   1
```

Sum	1	1	0	1	1	1
Carry	1	3	1	3	2	

i.e., $\mathbf{1011_2 \times 101_2 = 110\,111_2}$

For binary mixed numbers, the position of the binary point is ignored in the first instance, multiplication by repeated addition is performed as shown above and the binary point is inserted after multiplication, (see para 18).

Thus: $101.1_2 \times 1.01_2$ is treated as $1011_2 \times 101_2$ in the first instance,
i.e. $1011_2 \times 101_2 = 110\,111_2$, (see above) and
$101.1_2 \times 1.01_2 = 110.111_2$, (from para. 18).

20　When dividing number A by number B, giving C and remainder D, i.e. $\dfrac{A}{B} = C$, remainder D, A is called the **dividend**, B the **divisor** and C the **quotient**. To perform $51_{10} \div 9_{10}$ in binary, a similar procedure to that for long division of denary numbers is used.

$$51_{10} = 110\,011_2$$
$$9_{10} = 1001_2$$

```
                                        1   0   1   (quotient)
(divisor)  1   0   0   1  )  1  ⁰1̸  ¹0  ²0  1   1   (dividend

                             0   0   0   1
                             ───────────────
                             0   0   1   1   1   1
                                     1   0   0   1
                                     ───────────────
                                         1   1   0
```

Thus $\dfrac{110\,011_2}{1001_2} = 101_2$, remainder 110_2.

(Complementary addition may be used as an alternative to direct binary subtraction.)

21　Binary mixed numbers can be dealt with by expressing them as whole binary numbers. Thus:

$$\frac{10\,110.01_2}{1010.1_2} = \frac{10\,110\,01_2}{1010\,10_2}, \text{ and dividing as shown in para 20,}$$
gives:

```
                                    1   0
       1 0 1 0 1 0  ) 1  0  1  ⁰1̸  ²0̸  0  1
                      1  0  1  0  1   0
                      ─────────────────────
                      0  0  0  0  1   0  1
```

19

i.e. $\dfrac{1\,011\,001}{101\,010} = 10_2 + \dfrac{101_2}{101\,010_2}$

$$= 10_2 + \dfrac{1.01_2}{1010.1_2}$$

$$= 10_2, \text{ reaminder } 1.01_2$$

22 Binary division may be performed by repeated addition of the divisor. For example:

$$\dfrac{12_{10}}{3_{10}} = \dfrac{1100_2}{11_2}$$

Repeated addition of the divisor gives:

```
      11      once
      11      twice
    ————
     110
      11      three times
    ————
    1001
      11      four times
    ————
    1100      i.e. the dividend
    ————
```

Thus, $\dfrac{1100_2}{11_2} = \mathbf{100_2}$

Similarly, $\dfrac{1110_2}{11_2}$ gives:

```
      11      once
      11      twice
    ————
     110
      11      three times
    ————
    1001
      11      four times
    ————
    1100
      11      five times
    ————
    1111      which is larger than the dividend.
    ————
```

The number 1100, (the four times product) is larger than 1110_2 by 10_2.

Thus, $\dfrac{1110_2}{11_2} = 100_2$, remainder 10_2.

Coding

23 The natural binary system of numbers requires a large number of bits to depict a denary number having several significant figures. For example, the denary number 3 418 is the binary number 110101011010 and this binary number gives no real idea of the size of the number. Also, registers for storing bits in calculators, computers and microprocessors are usually of uniform size, containing room to store 4 or 8 or 16 bits. For these reasons, various **codes** are used in practice, rather than the natural binary system of numbers.

24 It is usual, when expressing a denary number as a binary number, to use a **decade system**. Each decade contains four bits and is used to represent a denary number from 0–9. The units decade is used to represent denary numbers from 0–9, the ten's decade is used for denary numbers from 10–99, and so on. This system of coding is called the **binary-coded-decimal system**, usually abbreviated to B.C.D. system.

Thus 2719_{10} encoded as a binary-coded-decimal number is given by:

$$2719_{10} = 0010\ 0111\ 0001\ 1001$$

Conversely, the B.C.D. number 1000 0110 0101 0011 expressed as a denary number is 8653_{10}.

25 Another code in common use is the 'excess-three code', witten as the **'X's 3 code'**, and when used as a BCD is called the **'X's 3 BCD'**. This code is formed by adding the binary number 0011 to the natural binary code. Thus the X's 3 binary numbers equivalent to the denary numbers 0, 1, 2, ... are 0011, 0100, 0101, ... The advantage of this code is that there is no binary number 0000, which may occur in the event of electrical failure to a system. Thus, to encode 5208_{10} as an excess-three, binary-coded-decimal:

Denary Number	Natural Binary Number	X's 3 Number
5	0101	$0101 + 0011 = 1000$
2	0010	$0010 + 0011 = 0101$
0	0000	$0000 + 0011 = 0011$
8	1000	$1000 + 0011 = 1011$

i.e. $5208_{10} = 1000\,0101\,0011\,1011$, when expressed as an X's 3
BCD.

26   During the transmission of binary codes, errors can occur due
to human errors and due to the signal being distorted in the
transmitting medium. When this occurs, a 0 in the signal is
received as a 1 and vice-versa.

Error detecting codes may be used, giving an indication of an
error having been made. The simplest of these error detecting
codes uses five bits instead of four, four bits containing the data
and the fifth bit being used for checking purposes, and called a
parity bit. Two such codes are the odd parity system and the
even parity system.

The odd parity system

27   The sum of the 1's in any group of five bits must add up to
an odd decimal number, the fifth parity bit being selected to meet
this requirement. Thus, 0100, 0110, 0111, 0000, 1111 is transmitted
as:

Information	Odd parity bit	Decimal sum
0100	0	1
0110	1	3
0111	0	3
0000	1	1
1111	1	5

i.e., 01000, 01101, 01110, 00001, 11111. If the five bits received do
not contain an odd number of 1's then the recipient knows an
error has been made.

Thus, for example, to encode 2976_{10} into a binary-coded-
decimal with odd parity:

Denary Number	BCD	Odd Parity Bit
2	0010	0
9	1001	1
7	0111	0
6	0110	1

i.e. $2976_{10} = 00100\,10011\,01110\,01101$ when expressed as a BCD
with odd parity.

The even parity system

28   This is similar to the odd parity system except that the
decimal sum of the five bits must be an even number. Thus 0100,
0110, 0111, 0000, 1111 is transmitted as 01001, 01100, 01111,

00000, 11110. If the five bits are received and the sum of the 1's is not an even decimal number, then the recipient knows an error has been made.

Thus, for example, to encode 4725_{10} into an excess-three, binary-coded-decimal with even parity:

Denary number	BCD	X's 3 BCD	Even parity bit
4	0100	0111	1
7	0111	1010	0
2	0010	0101	0
5	0101	1000	1

i.e 4725_{10} = **01111 10100 01010 10001 when expressed as an X's 3 BCD with even parity.**

29 Another error detecting code is called the **two-out-of-five code** in which five bits are used and the denary numbers 0 to 9 are each represented by two 1's and three 0's as shown below:

Two-out-of-five code	Corresponding denary number
11000	0
00011	1
00101	2
00110	3
01001	4
01010	5
01100	6
10001	7
10010	8
10100	9

If any set of five bits received contains more or less than two 1's then an error exists.

Thus, for example, to encode 5728_{10} into a two-out-of-five code:

Denary number	Two-out-of-five code
5	01010
7	10001
2	00101
8	10010

Thus 5728_{10} = **01010 10001 00101 10010 when expressed as a two-out-of-five code**.

3 Algebra

1   **Algebra** is that part of mathematics in which the relations and properties of numbers are investigated by means of general symbols. For example, the area of a rectangle is found by multiplying the length by the breadth; this is expressed algebraically as $A = l \times b$, where A represents the area, l the length and b the breadth.

The basic laws of algebra

2   Let a, b, c and d represent any four numbers. Then:

(i) $a + (b + c) = (a + b) + c$
(ii) $a(bc) = (ab)c$
(iii) $a + b = b + a$
(iv) $ab = ba$
(v) $a(b + c) = ab + ac$
(vi) $\dfrac{a+b}{c} = \dfrac{a}{c} + \dfrac{b}{c}$
(vii) $(a + b)(c + d) = ac + ad + bc + bd$

Thus, for example, from (v), $3(2 + 5) = 3 \times 2 + 3 \times 5 = 6 + 15 = 21$, from (vi), $\dfrac{3+4}{5} = \dfrac{3}{5} + \dfrac{4}{5}$ and from (vii),

$(2x + 3y)(a - b) = 2ax - 2bx + 3ay - 3by.$

Laws of indices

3   (i) $a^m \times a^n = a^{m+n}$. For example, $2^3 \times 2^4 = 2^{3+4} = 2^7 = 128$.

(ii) $\dfrac{a^m}{a^n} = a^{m-n}$.   For example, $\dfrac{3^5}{3^2} = 3^{5-2} = 3^3 = 27$.

(iii) $(a^m)^n = a^{mn}$.   For example, $(2^3)^4 = 2^{3 \times 4} = 2^{12} = 4096$.

(iv) $a^{m/n} = \sqrt[n]{a^m}$.   For example, $8^{2/3} = \sqrt[3]{8^2} = 2^2 = 4$.

(v) $a^{-n} = \dfrac{1}{a^n}$.   For example, $3^{-2} = \dfrac{1}{3^2} = \dfrac{1}{9}$.

(vi) $a^0 = 1$.   For example, $17^0 = 1$.

4   When two or more terms in an algebraic expression contain a common factor, then this factor can be shown outside of a bracket. For example, $ab + ac = a(b + c)$, which is simply the reverse of (v) in para 2, and $6px + 2py - 4pz = 2p(3x + y - 2z)$. This process is called **factorisation**. Similarly, since $3ab$ is common to all the terms in the expression $3a^2b - 6ab^2 + 15ab$, the expression factorises as $3ab(a - 2b + 5)$.

Direct and inverse proportionality

5   (i) An expression such as $y = 3x$ contains two variables. For every value of x there is a corresponding value of y. The variable x is called the **independent variable** and y is called the **dependent variable**.

(ii) When an increase or decrease in an independent variable leads to an increase or decrease of the same proportion in the dependent variable this is termed **direct proportion**. If $y = 3x$ then y is directly proportional to x, which may be written as $y \propto x$ or $y = kx$, where k is called the **coefficient of proportionality** (in this case, k being equal to 3.) Thus if y is directly proportional to x, and $y = 37.5$ when $x = 2.5$, then $y \propto x$, i.e. $y = kx$ and $37.5 = k(2.5)$, from which the constant of proportionality,

$$k = \frac{37.5}{2.5} = 15.$$

Thus the value of y when $x = 6$ is given by $y = kx = (15)(6) = 90$. Examples of laws involving direct proportion include Hooke's law, Charles' law and Ohm's law.

(iii) When an increase in an independent variable leads to a decrease of the same proportion in the dependent variable (or vice versa) this is termed **inverse proportion**. If y is inversely proportional to x then

$$y \propto \frac{1}{x} \text{ or } y = \frac{k}{x}.$$

Alternatively, $k = xy$, that is, for inverse proportionality the product of the variables is constant. An example of a law involving inverse proportion is Boyles law. Thus if volume V is inversely proportional to pressure p and $V = 0.08$ when $p = 1.5 \times 10^6$, then $V \propto \frac{1}{p}$, i.e. $V = \frac{k}{p}$, from which the constant of proportionality,

$$k = Vp = (0.08)(1.5 \times 10^6) = 12 \times 10^4.$$

Thus the volume when the pressure changes to 4×10^6 is given by:

$$V = \frac{k}{p} = \frac{12 \times 10^4}{4 \times 10^6} = 3 \times 10^{-2} \text{ or } 0.03.$$

Equations

6   An equation is a statement that two quantities are equal. To **'solve an equation'** means 'to find the value of the unknown'. The value of the unknown is called the **root** of the equation.

7    $(2x - 3)$ is an example of an **algebraic expression**, whereas $2x - 3 = 1$ is an example of an equation (i.e. it contains an 'equals' sign).

8   An **identity** is a relationship which is true for all values of the unknown, whereas an equation is only true for particular values of the unknown. For example, $2x - 3 = 1$ is an equation since it is only true when $x = 2$, whereas $3x \equiv 8x - 5x$ is an identity since it is true for **all** values of x. (*Note*: '$\equiv$' means 'is identical to'.)

9   A **linear equation** is one in which an unknown quantity is raised to the power 1. $3x + 2 = 0$ is an example of a linear equation. Any arithmetic operation may be applied to an equation **as long as the equality of the equation is maintained**.

For example, if 2 is subtracted from **both** sides of the equation $3x + 2 = 0$,

then $3x = -2$.

Dividing **both** sides by 3 gives: $x = -\dfrac{2}{3}$. Thus $x = -\dfrac{2}{3}$ is the root of the linear equation $3x + 2 = 0$.

To solve $3(x - 2) = 9$, the brackets are firstly removed.

Thus $3x - 6 = 9$.

Then $3x = 9 + 6 = 15$,

and $x = \dfrac{15}{3} = \mathbf{5}$.

To solve $\dfrac{3}{x} = \dfrac{4}{5}$, each side of the equation is multiplied by the LCM of the denominator, $5x$.

Hence $5x\left(\dfrac{3}{x}\right) = 5x\left(\dfrac{4}{5}\right)$

Cancelling gives: $(5)(3) = (x)(4)$

i.e. $\qquad 15 = 4x,$

from which, $\qquad x = \dfrac{15}{4} = 3\dfrac{3}{4}.$

10   A **quadratic equation** is one in which the highest power of the unknown quantity is 2. For example, $x^2 - 3x + 1 = 0$ is a quadratic equation. There are **four methods of solving quadratic equations**. These are:

(i) by factorisation (where possible),

(ii) by 'completing the square',

(iii) by using the 'quadratic formula', and

(iv) graphically, (see page 112).

11   Multiplying out $(2x+1)(x-3)$ gives $2x^2 - 6x + x - 3$, i.e., $2x^2 - 5x - 3$. The reverse process of moving from $2x^2 - 5x - 3$ to $(2x+1)(x-3)$ is called **factorising**. If the quadratic expression can be factorised this provides the simplest method of solving a quadratic equation. For example, if $2x^2 - 5x - 3 = 0$, then, by factorising:

$$(2x + 1)(x - 3) = 0.$$

Hence, either $(2x+1) = 0$, i.e., $x = -\dfrac{1}{2}$

or $(x - 3) = 0$, i.e., $x = 3$.

The technique of factorising is often one of 'trial and error'.

12   (i) Let the general quadratic equation be $ax^2 + bx + c = 0$, where a, b and c are constants.

Dividing throughout by a gives: $x^2 + \dfrac{b}{a}x + \dfrac{c}{a} = 0,$

and rearranging gives: $\qquad x^2 + \dfrac{b}{a}x = -\dfrac{c}{a}.$

Adding to each side of this equation the square of half the coefficient of the term in x to make the left hand side a perfect square gives:

$$x^2 + \dfrac{b}{a}x + \left(\dfrac{b}{2a}\right)^2 = \left(\dfrac{b}{2a}\right)^2 - \dfrac{c}{a}.$$

Rearranging gives: $\left(x + \dfrac{b}{2a}\right)^2 = \dfrac{b^2}{4a^2} - \dfrac{c}{a} = \dfrac{b^2 - 4ac}{4a^2}$

Taking the square root of both sides gives:

$$x + \dfrac{b}{2a} = \sqrt{\left(\dfrac{b^2 - 4ac}{4a^2}\right)} = \dfrac{\pm \sqrt{(b^2 - 4ac)}}{2a}$$

27

Hence, $x = -\dfrac{b}{2a} \pm \dfrac{\sqrt{(b^2 - 4ac)}}{2a}$

i.e. $x = \dfrac{-b \pm \sqrt{(b^2 - 4ac)}}{2a}$

This method of solution is called **'completing the square'**.

(ii) Summarising, if $ax^2 + bx + c = 0$, then

$$x = \frac{-b \pm \sqrt{(b^2 - 4ac)}}{2a}$$

This is known as the **quadratic formula**.

Thus to solve $3x^2 - 11x - 4 = 0$, $a = 3$, $b = -11$ and $c = -4$.

Hence $x = \dfrac{-(-11) \pm \sqrt{[(-11)^2 - 4(3)(-4)]}}{2(3)}$

$= \dfrac{11 \pm \sqrt{121 + 48]}}{6}$

$= \dfrac{11 \pm \sqrt{169}}{6} = \dfrac{11 \pm 13}{6} = \dfrac{11 + 13}{6}$ or $\dfrac{11 - 13}{6}$

Hence $x = \dfrac{24}{6} = \mathbf{4}$ or $x = -\dfrac{2}{6} = -\dfrac{\mathbf{1}}{\mathbf{3}}$

13 A **cubic equation** is one in which the highest power of the unknown quantity is 3. For example, $2x^3 + x^2 - x + 4 = 0$ is a cubic equation. Cubic equations may be solved (i) by iterative methods (see paras. 17 to 21) or (ii) by plotting graphs (see section 8, para. 21, page 115).

14 Equations which have to be solved together to find the unique values of the unknown quantities, which are true for each of the equations, are called **simultaneous equations**.

There are **three methods of solving a pair of linear simultaneous equations in two unknowns**. These are:

(i) by substitution,
(ii) by elimination,
(iii) graphically, (see page 109)

Thus, for example, to solve the simultaneous equations

$x + 2y = -1$ (1)
$4x - 3y = 18$ (2)

by substitution:

From equation (1), $x = -1 - 2y$.

Substituting $x = -1 - 2y$ into equation (2) gives

$$4(-1-2y)-3y=18$$
i.e. $-4-8y-3y=18$
$$-11y=18+4=22$$
from which, $y=\dfrac{22}{-11}=-2$

Substituting $y=-2$ into equation (1) gives
$$x+2(-2)=-1$$
from which, $x-4=-1$ and $x=3$

To solve the given simultaneous equations by elimination:
If equation (1) is multiplied throughout by 4, the coefficient of x will be the same as in equation (2) giving:

$$4x+8y=-4 \tag{3}$$

Subtracting equation (3) from equation (2) gives:

$$\begin{aligned} 4x-3y &= 18 \\ - \quad 4x+8y &= -4 \\ \hline 0-11y &= 22 \end{aligned}$$

(2)

(3)

Hence $y=\dfrac{22}{-11}=-2$ as before.

By substituting this value of y into equation (1) or (2) gives $x=3$ as before. The solution $x=3$, $y=-2$ is the only pair of values that satisfy both of the original equations.

15   There are two methods of solving simultaneous equations, one of which is linear and the other quadratic. These are: (i) by substitution, or (ii) graphically. (See section 8, para 20, page 113). For example, to solve by substitution,

$$y=2x^2-3x-4 \tag{1}$$

$$y=2-4x \tag{2}$$

the two equations are equated,
i.e. $\qquad 2x^2-3x-4=2-4x$
from which $\qquad 2x^2+x-6=0$.
Factorising gives:
$$(2x-3)(x+2)=0$$
from which: $x=\dfrac{3}{2}$ and $x=-2$.

In equation (2), when $x=\dfrac{3}{2}$, $y=-4$ and when $x=-2$, $y=10$

which may be checked in equation (1).

16   (i) The statement $v=u+at$ is said to be a **formula** for v in

terms of u, a and t. v, u, a and t are called **symbols**. The single term on the left hand side of the equation, v, is called the **subject of the formula.**

(ii) Provided values are given for all the symbols in a formula except one, the remaining symbol can be made the subject of the formula and may be evaluated by using tables, or calculators.

(iii) When a symbol other than the subject is required to be calculated it is usual to rearrange the formula to make a new subject. This rearranging process is called **transposing the formula** or **transposition**.

(iv) The rules for transposition of formulae are the same as those used for the solution of simple equations (see para 9) — i.e. the equality of an equation must be maintained.

Thus if $v = f\lambda$, then to make λ the subject of the formula both sides of the equation are divided by f.

i.e., $\dfrac{v}{f} = \lambda$

Similarly, to transpose $R = \dfrac{\rho l}{a}$ for a, initially multiply both sides of the equation by a giving $Ra = \rho l$, and then dividing both sides by R, giving $\boldsymbol{a = \dfrac{\rho l}{R}}$.

Also, to transpose $v = u + \dfrac{ft}{m}$ for f, firstly rearrange to obtain the term in f on its own on one side of the equation, i.e subtract u from both sides, giving $v - u = \dfrac{ft}{m}$.

Multiplying each side by m gives: $m(v - u) = ft$

and dividing both sides by t gives: $\boldsymbol{\dfrac{m}{t}(v - u) = f.}$

Iterative methods

17   Many equations can only be solved graphically or by methods of successive approximations to the roots, called iterative methods. Two methods of successive approximations are (i) an algebraic method, and (ii) by using the Newton-Raphson formula.

18   Both successive approximation methods rely on a reasonably

good first estimate of the value of a root being made. One way of doing this is to sketch a graph of the function, say, $y = f(x)$, and determine the approximate value of roots from the points where the graph cuts the x-axis. Another way is by using a functional rotation method. This method uses the property that the value of the graph of $f(x) = 0$ changes sign for values of x just before and just after the value of a root. For example, for the equation

$$4x^2 - 6x - 7 = 0,$$

let $f(x) = 4x^2 - 6x - 7$
Then $f(0) = -7$
and $f(-1) = 4(-1)^2 - 6(-1) - 7 = 3$.

Since the value of $f(x)$ changes from -7 to $+3$, a root of the equation must be between $x = 0$ and $x = -1$. The difference between -7 and $+3$ is 10 thus a reasonable first approximation is $\dfrac{-7}{10}$ from 0, i.e., -0.7 $\left(\text{or } \dfrac{+3}{10} \text{ back from } -1, \text{ i.e., } -0.7\right)$.

An algebraic method of successive approximations

19 This method can be used to solve equations of the form:

$$a + bx + cx^2 + dx^3 + \ldots = 0,$$

where $a, b, c, d, \ldots$ are constants.
Procedure:

First approximation
  (i) Using a graphical or the functional notation method, (see para 18), determine an approximate value of the root required, say x_1.

Second approximation
  (ii) Let the true value of the root be $(x_1 + \delta_1)$.
  (iii) Determine x_2 the approximate value of $(x_1 + \delta_1)$ by determining the value of $f(x_1 + \delta_1) = 0$, but neglecting terms containing products of δ_1.

Third approximation
  (iv) Let the true value of the root be $(x_2 + \delta_2)$.
  (v) Determine x_3, the approximate value of $(x_2 + \delta_2)$ by determining the value of $f(x_2 + \delta_2) = 0$, but neglecting terms containing products of δ_2.

31

(vi) The fourth and higher approximations are obtained in a similar way.

20   Using the techniques given in para 19(b) to (f), it is possible to continue getting values nearer and nearer to the required root. The procedure is repeated until the value of the required root does not change on two consecutive approximations, when expressed to the required degree of accuracy. For example, to determine the smallest positive root of the equation $3x^3 - 10x^2 + 4x + 7 = 0$, correct to 3 significant figures, the functional notation method is used initially to find the value of the first approximation.

$$f(x) = 3x^3 - 10x^2 + 4x + 7$$
$$f(0) = 3(0)^3 - 10(0)^2 + 4(0) + 7 = 7$$
$$f(1) = 3(1)^3 - 10(1)^2 + 4(1) + 7 = 4$$
$$f(2) = 3(2)^3 - 10(2)^2 + 4(2) + 7 = -1$$

Following the procedure given in para 19:

First approximation
(i) Let the first approximation be such that it divides the interval 1 to 2 in the ratio of 4 to -1, i.e., let x_1 be 1.8.

Second approximation
(ii) Let the true value of the root, x_2, be $(x_1 + \delta_1)$.
(iii) Let $f(x_1 + \delta_1) = 0$, then since $x_1 = 1.8$,
$$3(1.8 + \delta_1)^3 - 10(1.8 + \delta_1)^2 + 4(1.8 + \delta) + 7 = 0$$

Neglecting terms containing products of δ_1 and using the binomial series (see section 4, page 41), gives:

$$3[1.8^3 + 3(1.8)^2\delta_1] - 10[1.8^2 + (2)(1.8)\delta_1] + 4(1.8 + \delta_1) + 7 \simeq 0$$
$$3(5.832 + 9.72\delta_1) - 32.4 - 36\delta_1 + 7.2 + 4\delta_1 + 7 \simeq 0$$
$$17.496 + 29.16\delta_1 - 32.4 - 36\delta_1 + 7.2 + 4\delta_1 + 7 \simeq 0$$
$$d_1 \simeq \frac{-17.496 + 32.4 - 7.2 - 7}{29.16 - 36 + 4} \simeq \frac{0.704}{2.84} \simeq -0.25$$

Thus $x_2 \simeq 1.8 - 0.25 = 1.55$

Third approximation
(iv) Let the true value of the root, x_3, be $(x_2 + \delta_2)$.
(v) Let $f(x_2 + \delta_2) = 0$, then since $x_2 = 1.55$,
$$3(1.55 + \delta_2)^3 - 10(1.55 + \delta_2)^2 + 4(1.55 + \delta_2) + 7 = 0$$

Neglecting terms containing products of δ_2, gives:

$$11.17 + 21.62\delta_2 - 24.03 - 31\delta_2 + 6.2 + 4\delta_2 + 7 \approx 0$$
$$\delta_2 \approx \frac{-11.17 + 24.03 - 6.2 - 7}{21.62 - 31 + 4} \approx \frac{-0.34}{-5.38} = 0.063$$

Thus $x_3 \approx 1.55 + 0.063 = 1.613$.

(vi) Values of x_4 and x_5 are found in a similar way.

$f(x_3 + \delta_3) =$

$3(1.613 + \delta_3)^3 - 10(1.613 + \delta_3)^2 + 4(1.613 + \delta_3) + 7 = 0$
giving $\delta_3 \approx 0.005$ and $x_4 \approx 1.618$ i.e., 1.62 correct to 3 significant figures.

$f(x_4 + \delta_4) = 3(1.618 + \delta_4)^3 - 10(1.618 + \delta_4)^2 + 4(1.618 + \delta_4) + 7 = 0$
giving $\delta_4 = 0$, correct to 4 significant figures and
$x_5 \approx 1.62$, correct to 3 significant figures.

Since x_4 and x_5 are the same when expressed to the required degree of accuracy, then, the required root is **1.62**, correct to 3 significant figures.

(*Note on accuracy and errors.* Depending on the accuracy of evaluating the $f(x + \delta)$ terms, one or two iterations (i.e. successive approximations) might be saved. However, it is not usual to work to more than about 4 significant figures accuracy in this type of calculation. If a small error is made in calculations, the only likely effect is to increase the number of iterations.)

Newton's method

21   The Newton-Raphson formula, often just referred to as **Newton's method** may be stated as follows:

if r_1 is the approximate value of a real root of the equation $f(x) = 0$, then a closer approximation to the root, r_2 is given by:

$$r_2 = r_1 - \frac{f(r_1)}{f'(r_1)}$$

(If, as occasionally happens, the successive approximations of a root do not converge towards the value of the root, a new value of r_1 should be selected so that $f(r_1)$ has the same sign as $f''(r_1)$). The advantages of Newton's method over the algebraic method of successive approximations is that it can be used for any type of mathematical equation, (i.e. ones containing trigonometric, exponential, logarithmic, hyperbolic and algebraic functions), and it is usually easier to apply than the algebraic method.

Thus, for example, to find the root of the equation

$$x^2 - 3 \sin x + 2 \ln(x + 1) = 3.5,$$

correct to 3 significant figures:
$f(x) = x^2 - 3 \sin x + 2 \ln(x + 1) - 3.5$
$f(0) = -3.5$

$f(1) = 1 - 3 \sin 1 + 2 \ln 2 - 3.5 = -3.6381$ (*Note*: 'sin 1' means 'the sine of 1 radian')

$f(2) = 4 - 3 \sin 2 + 2 \ln 3 - 3.5 = -0.0307$

$f(3) = 9 - 3 \sin 3 + 2 \ln 4 - 3.5 = 7.8492$.

Hence, let the first approximation, $r_1 = 2$.

Newton's formula states that $r_2 = r_1 - \dfrac{f(r_1)}{f'(r_1)}$

where r_1 is a first approximation to the root and r_2 is a better approximation to the root.

Hence, $f(r_1) = -0.0307$.

$$f'(x) = 2x - 3 \cos x + \frac{2}{x+1} \quad \text{(see section 12)}$$

$$f'(r_1) = f''(2) = 2(2) - 3 \cos 2 + \frac{2}{3}$$

$$= 4 + 1.2484 + 0.6667 = 5.9151.$$

Hence, $r_2 = r_1 - \dfrac{f(r_1)}{f'(r_1)} = 2 - \dfrac{-0.0307}{5.9151}$

$\qquad = 2.005$ or 2.01, correct to 3 significant figures.

A still better approximation to the root, r_3, is given by:

$$r_3 = r_2 - \frac{f(r_2)}{f'(r_2)}$$

$$= 2.005 - \frac{[(2.005)^2 - 3 \sin 2.005 + 2 \ln 3.005 - 3.5]}{\left[2(2.005) - 3 \cos 2.005 + \dfrac{2}{2.005 + 1} \right]}$$

$$= 2.005 - \frac{(-0.0010)}{5.938} = 2.005 + 0.00017$$

i.e. $r_3 = 2.01$, correct to 3 significant figures.

Since the values of r_2 and r_3 are the same when expressed to the required degree of accuracy, then the required root is **2.01**, correct to 3 significant figures.

Partial fractions

22   By algebraic addition,

$$\frac{1}{x-2} + \frac{3}{x+1} = \frac{(x+1) + 3(x-2)}{(x-2)(x+1)} = \frac{4x-5}{x^2 - x - 2}$$

The reverse process of moving from

$$\frac{4x-5}{x^2 - x - 2} \quad \text{to} \quad \frac{1}{x-2} + \frac{3}{x+1}$$

is called resolving into **partial fractions**.

In order to resolve an algebraic expression into partial fractions:

  (i) the denominator must factorise (in the above example, $x^2 - x - 2$ factorises as $(x-2)(x+1)$), and

  (ii) the numerator must be at least one degree less than the denominator (in the above example $(4x-5)$ is of degree 1 since the highest powered x term is x^1, and $(x^2 - x - 2)$ is of degree 2).

When the degree of the numerator is equal to or higher than the degree of the denominator, the numerator must be divided by the denominator until the remainder is of less degree than the denominator.

There are basically three types of partial fraction and the form of partial fraction used is summarised below, where $f(x)$ is assumed to be of less degree than the relevant denominator and A, B, and C are constants to be determined.

Denominator containing	Expression	Form of partial fraction
Type 1 Linear factors	$\dfrac{f(x)}{(x+a)(x-b)(x+c)}$	$\dfrac{A}{(x+a)} + \dfrac{B}{(x-b)} + \dfrac{C}{(x+c)}$
Type 2 Repeat linear factors	$\dfrac{f(x)}{(x+a)^3}$	$\dfrac{A}{(x+a)} + \dfrac{B}{(x+a)^2} + \dfrac{C}{(x+a)^3}$
Type 3 Quadratic factors	$\dfrac{f(x)}{(ax^2+bx+c)(x+d)}$	$\dfrac{Ax+B}{(ax^2+bx+c)} + \dfrac{C}{(x+d)}$

(In the latter type, $ax^2 + bx + c$ is a quadratic expression which does not factorise without containing surds or imaginary terms.)

Thus, for example, to resolve $\dfrac{11-3x}{x^2+2x-3}$ into partial fractions it is firstly recognised that the denominator factorises as $(x-1)(x+3)$ and the numerator is of less degree than the denominator.

Thus $\dfrac{11-3x}{x^2+2x-3}$ may be resolved into partial fractions.

Let $\dfrac{11-3x}{(x-1)(x+3)} = \dfrac{A}{(x-1)} + \dfrac{B}{(x+3)}$, where A and B are

constants to be determined, i.e., $\dfrac{11-3x}{(x-1)(x+3)} \equiv \dfrac{A(x+3)+B(x-1)}{(x-1)(x+3)}$,

by algebraic addition.

Since the denominators are the same on each side of the identity then the numerators are equal to each other. Thus,
$11-3x = A(x+3)+B(x-1)$.
To determine constants A and B, values of x are chosen to make the term in A or B equal to zero. When $x=1$, then
$11 - 3(1) = A(1+3) + B(0)$

   i.e. $8 = 4A$
   i.e. $A = 2$

When $x = -3$, then $11 - 3(-3) = A(0) + B(-3-1)$

   i.e. $20 = -4B$
   i.e. $B = -5$

Thus $\dfrac{11-3x}{x^2+2x-3} = \dfrac{2}{(x-1)} + \dfrac{-5}{(x+3)} \equiv \dfrac{\mathbf{2}}{(\mathbf{x-1})} - \dfrac{\mathbf{5}}{(\mathbf{x+3})}$

$\left[\text{Check} : \dfrac{2}{x-1} - \dfrac{5}{x+3} \equiv \dfrac{2(x+3)-5(x-1)}{(x-1)(x+3)} = \dfrac{11-3x}{x^2+2x-3} \right]$

Similarly, when resolving $\dfrac{2x+3}{(x-2)^2}$ into partial fractions it is

recognised that the denominator contains a repeated linear factor, $(x-2)^2$.

   Let $\dfrac{2x+3}{(x-2)^2} \equiv \dfrac{A}{(x-2)} + \dfrac{B}{(x-2)^2} = \dfrac{A(x-2)+B}{(x-2)^2}$

Equating the numerators gives: $2x+3 \equiv A(x-2)+B$
Let $x=2$. Then $7 = A(0) + B$ i.e. $B = 7$
$2x+3 \equiv A(x-2)+B \equiv Ax-2A+B$.

Since an identity is true for all values of the unknown, the coefficients of similar terms may be equated.
Hence, equating the coefficients of x gives: $2 = A$
Also, as a check, equating the constant terms gives: $3 = -2A + B$
When $A=2$, $B=7$, R.H.S. $= -2(2) + 7 = 3 = $ L.H.S.
Hence,

$$\dfrac{\mathbf{2x+3}}{\mathbf{(x-2)^2}} = \dfrac{\mathbf{2}}{\mathbf{(x-2)}} + \dfrac{\mathbf{7}}{\mathbf{(x-2)^2}}$$

To resolve $\dfrac{3+6x+4x^2-2x^3}{x^2(x^2+3)}$ into partial fractions, it is

recognised that the denominator is a combination of a quadratic factor, (x^2+3), which does not factorise without introducing imaginary surd terms, and repeated linear factors — terms such as x^2 being treated as $(x+0)^2$. Let

$$\frac{3+6x+4x^2-2x^3}{x^2(x^2+3)} \equiv \frac{A}{x} + \frac{B}{x^2} - \frac{Cx+D}{(x^2+3)}$$
$$\equiv \frac{Ax(x^2+3)+B(x^2+3)+(Cx+D)x^2}{x^2(x^2+3)}$$

Equating the numerators gives:

$$3+6x+4x^2-2x^3 \equiv Ax(x^2+3)+B(x^2+3)+(Cx+D)x^2$$
$$\equiv Ax^3+3Ax+Bx^2+3B+Cx^3+Dx^2$$

Let $x=0$. Then $3=3B$

    i.e. $B=1$

Equating the coefficients of x^3 terms gives: $-2=A+C$ (1)
Equating the coefficients of x^2 terms gives: $4=B+D$
Since $B=1$, $D=3$.
Equating the coefficients of x terms gives: $6=3A$
i.e. $A=2$
From equation (1), since $A=2$, $C=-4$
Hence

$$\frac{3+6x+4x^2-2x^3}{x^2(x^2+3)}=\frac{2}{x}+\frac{1}{x^2}+\frac{-4x+3}{x^2+3}\equiv\frac{\mathbf{2}}{x}+\frac{\mathbf{1}}{x^2}+\frac{\mathbf{3-4x}}{x^2+3}$$

26   Resolving an algebraic expression into partial fractions is used as a preliminary to integrating certain functions. (See para 12, page 193).

Exponential functions

27   An exponential function is one which contains e^x, e being a constant called the **exponent** and having an approximate value of 2.7183. The exponent arises from the natural laws of growth and decay and is used as a base of natural or Naperian logarithms.

28   **The natural laws of growth and decay** are of the form $y=Ae^{kx}$, where A and k are constants. The natural laws occur frequently in engineering and science and examples of quantities related by a natural law include:

(i) Linear expansion $l = l_0 e^{\alpha\theta}$

(ii) Change in electrical resistance with temperature $R_0 = R_0 e^{\alpha\theta}$

(iii) Tension in belts $T_1 = T_0 e^{\mu\alpha}$

(iv) Newton's law of cooling $\theta = \theta_0 e^{-kt}$

(v) Biological growth $y = y_0 e^{kt}$

(vi) Discharge of a capacitor $q = Q e^{-(t/CR)}$

(vii) Atmospheric pressure $p = p_0 e^{-(h/e)}$

(viii) Radioactive decay $\mathcal{N} = \mathcal{N}_0 e^{-\lambda t}$

(ix) Decay of current in an inductive circuit $i = I e^{-(Rt/L)}$

29   The **value of** e^x may be determined by using

(i) A calculator which possesses an 'e^x' function,

(ii) The power series $e^x = 1 + x + \dfrac{x^2}{2!} + \dfrac{x^3}{3!} + \ldots$ (where $3!$ is 'factorial 3' and means $3 \times 2 \times 1$),

(iii) Naperian logarithms, or

(iv) 4-figure tables of exponential functions.

Examples include $e^{0.36} = 1.4333$, $e^{-0.47} = 0.6250$

and $e^{-2.6} = 0.0743$, each correct to 4 decimal places,

and $e^{1.7629} = 5.829318$, correct to 7 significant figures.

30   (i) *Figure 3.1* shows graphs of $y = e^x$ and $y = e^{-x}$.

(ii) A graph of $y = 5e^{(1/2)x}$ is shown in *Figure 3.2*. This is obtained by drawing up a table of values of x and determining the corresponding values of y.

        The gradient of the curve at any point, $\dfrac{dy}{dx}$, is obtained by drawing a tangent to the curve at that point and measuring the gradient of the tangent.

        For example, when $x = 0$, $y = 5$ and $\dfrac{dy}{dx} = \dfrac{BC}{AB} = \dfrac{(6.2 - 3.7)}{1} = 2.5$

and when $x = 2$, $y = 13.6$ and $\dfrac{dy}{dx} = \dfrac{EF}{DE} = \dfrac{(16.8 - 10)}{1} = 6.8$

        These two results each show that $\dfrac{dy}{dx} = \dfrac{1}{2}y$, and further determinations of the gradients of $y = 5e^{(1/2)x}$ would give the same result for each. In general, for all natural growth and decay laws of the form $y = Ae^{kx}$, where k is a positive constant for growth laws (as in *Figure 3.2*) and a negative constant for decay occurs, $\dfrac{dy}{dx} = ky$,

i.e., **the rate of change of the variable, y, is proportional to the variable itself**.

38

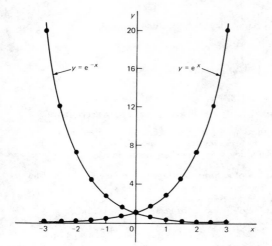

Figure 3.1

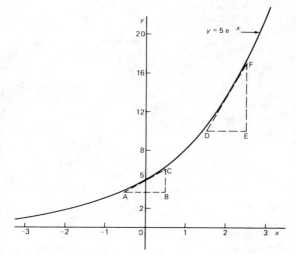

Figure 3.2

(iii) For any natural law of growth and decay of the form $\dfrac{dy}{dx} = ky$, the solution is always $y = Ae^{kx}$. Thus, for example,

$$\text{if } \frac{dQ}{dt} = \left(-\frac{1}{CR} \right) Q \qquad\qquad \text{then } Q = Ae^{[-(1/CR)]t}$$

Hyperbolic or Naperian logarithms

31   Logarithms having a base of e are called **hyperbolic**, **Naperian** or **natural logarithms** and the Naperian logarithm of x is written as $\log_e x$, or more commonly, $\ln x$.

32   The **value of a Naperian logarithm** may be obtained by using:
(i) A calculator possessing an '$\ln x$' function,
(ii) The change of base rule for logarithms, which states,

$$\log_a y = \frac{\log_b y}{\log_b a}, \text{ from which, } \ln y = \frac{\lg y}{\lg e} = \frac{\lg y}{0.4343} = 2.3026 \lg y;$$

or

(iii) 4-figure Naperian logarithm tables.
For example, $\ln 3.478 = 1.2465$
$\qquad\qquad\quad \ln 469.2 = 6.1510$
and $\qquad\quad \ln 0.07314 = -2.6154$, each correct to 4 decimal places.

4 Series

Binomial theorem

1   A **binomial expression** is one which contains two terms connected by a plus or minus sign. Thus $(p+q)$, $(a-x)^2$, $(2x+y)^3$ are examples of binomial expressions.

2   Expanding $(a+x)^n$ for integer values of n from 0 to 6 gives the following results:

$$
\begin{aligned}
(a+x)^0 &= &1 \\
(a+x)^1 &= &a+x \\
(a+x)^2 = (a+x)(a+x) &= &a^2 + 2ax + x^2 \\
(a+x)^3 = (a+x)^2(a+x) &= &a^3 + 3a^2x + 3ax^2 + x^3 \\
(a+x)^4 = (a+x)^3(a+x) &= &a^4 + 4a^3x + 6a^2x^2 + 4ax^3 + x^4 \\
(a+x)^5 = (a+x)^4(a+x) &= &a^5 + 5a^4x + 10a^3x^2 + 10a^2x^3 + 5ax^4 + x^5 \\
(a+x)^6 = (a+x)^5(a+x) &= &a^6 + 6a^5x + 15a^4x^2 + 20a^3x^3 + 15a^2x^4 + 6ax^5 + x^6
\end{aligned}
$$

3   From the results of para 2 the following patterns emerge:

   (i) 'a' decreases in power moving from left to right.

   (ii) 'x' increases in power moving from left to right.

   (iii) **The coefficients of each term of the expansions are symmetrical about the middle coefficient when n is even and symmetrical about the two middle coefficients when n is odd.**

   (iv) The coefficients are shown separately in *Table 4.1* and this arrangement is known as **Pascal's triangle**. A coefficient of a term may be obtained by adding the two adjacent coefficients immediately above in the previous row. This is shown by the triangles in *Table 4.1*, where, for example, $1+3=4$, $10+5=15$, and so on.

   (v) Pascal's triangle method is used for expansion of the form $(a+x)^n$ for integer values of n less than about 8.

4   The **binomial theorem** is a formula for raising a binomial expression to any power without lengthy multiplication. The general binomial expansion of $(a+x)^n$ is given by:

$$
(a+n)^n = a^n + na^{n-1}x + \frac{n(n-1)}{2!}a^{n-2}x^2 + \frac{n(n-1)(n-2)}{3!}a^{n-3}x^3 + \ldots + x^n
$$

Table 4.1

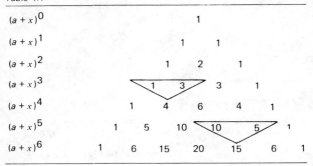

$(a+x)^0$							1						
$(a+x)^1$						1		1					
$(a+x)^2$					1		2		1				
$(a+x)^3$				1		3		3		1			
$(a+x)^4$			1		4		6		4		1		
$(a+x)^5$		1		5		10		10		5		1	
$(a+x)^6$	1		6		15		20		15		6		1

where 3! denotes $3 \times 2 \times 1$ and is termed 'factorial 3'. With the binomial theorem n may be a fraction, a decimal fraction or a positive or negative integer. For example,

$$(2a+3b)^5 = (2a)^5 + 5(2a)^4(3b) + \frac{(5)(4)}{2!}(2a)^3(3b)^2 + \frac{(5)(4)(3)}{3!}(2a)^2(3b)^3$$
$$+ \frac{(5)(4)(3)(2)}{4!}(2a)(3b)^4 + (3b)^5$$
$$= 32a^5 + 240a^4b + 720a^3b^2 + 1080a^2b^3 + 810ab^4 + 243b^5$$

5   The r'th term of the expansion $(a+x)^n$ is:

$$\frac{n(n-1)(n-2)\ldots \text{ to } (r-1) \text{ terms}}{(r-1)!} a^{n-(r-1)} x^{r-1}$$

For example, the 5'th term of $(3+x)^7$ is $\frac{(7)(6)(5)(4)}{(5-1)!} 3^{7-4} x^4$, since $r=5$ i.e. $(35)3^3 x^4 = \mathbf{945x^4}$

6   If $a=1$ in the binomial expansion of $(a+x)^n$ then:

$$(1+x)^n = 1 + nx + \frac{n(n-1)}{2!}x^2 + \frac{n(n-1)(n-2)}{3!}x^3 + \ldots,$$

which is valid for $-1 < x < 1$. When x is small compared with 1 then: $(1+x)^n \approx +nx$.

7   Binomial expansions may be used for numerical approximations, for calculations with small variations and in probability theory.

Maclaurin's theorem

8 Maclaurin's theorem states:

$$f(x) = f(0) + xf'(0) + \frac{x^2}{2!}f''(0) + \ldots$$

Some of the results obtained by applying Maclaurin's theorem to various functions are listed in paras 9 to 12.

Trigonometric series

9 $\sin x = x - \frac{x^3}{3!} + \frac{x^5}{5!} - \frac{x^7}{7!} + \ldots$ (valid for all values of x)

$\cos x = 1 - \frac{x^2}{2!} + \frac{x^4}{4!} - \frac{x^6}{6!} + \ldots$ (valid for all values of x)

Exponential series

10 $e^x = 1 + x + \frac{x^2}{2!} + \frac{x^3}{3!} + \ldots$ (valid for all values of x)

Logarithmic series

11 $\ln(1 + x) = x - \frac{x^2}{2} + \frac{x^3}{3} - \frac{x^4}{4} + \ldots$ (valid if $-1 < x \leqslant 1$)

Hyperbolic series

12 $\sinh x = x + \frac{x^3}{3!} + \frac{x^5}{5!} + \frac{x^7}{7!} + \ldots$ (valid for all values of x)

$\cosh x = 1 + \frac{x^2}{2!} + \frac{x^4}{4!} + \frac{x^6}{6!} + \ldots$ (valid for all values of x)

(See section 11, para 9, page 159)

Taylor's theorem

13 $f(a + h) = f(a) + hf'(a) + \frac{h^2}{2!}f''(a) + \ldots$

Some applications of Taylor's theorem include numerical differentiation, limits, small errors and the numerical solution of certain differential equations.

Arithmetical progressions

14 When a sequence has a constant difference between successive terms it is called an **arithmetical progression** (often abbreviated

to A.P.). Examples include (i) 1, 4, 7, 10, 13, ... where the common difference is 3, and (ii) a, $a+d$, $a+2d$, $a+3d$, ... where the common difference is d.

15   If the first term of an A.P. is 'a' and the common difference is 'd' then the n'th term is: $a+(n-1)d$

In example (i) of para 14, the 7'th term is given by $1+(7-1)3=19$, which may be readily checked.

16   The sum S of an A.P. can be obtained by multiplying the average of all the terms by the number of terms.

The average of all the terms $=\dfrac{a+l}{2}$, where 'a' is the first term and l is the last term.

i.e. $l=a+(n-1)d$, for n terms.

Hence the sum of n terms, $S_n = n\left(\dfrac{a+l}{2}\right) = \dfrac{n}{2}\{a + [a+(n-1)d]\}$

i.e. $S_n = \dfrac{n}{2}[2a+(n-1)d]$

For example, the sum of the first 7 terms of the series, 1, 4, 7, 10, 13, ... is given by

$$S_7 = \frac{7}{2}[2(1)+(7-1)3], \text{ since } a=1 \text{ and } d=3,$$

$$=\frac{7}{2}[2+18] = \frac{7}{2}(20) = \mathbf{70}.$$

Geometric progressions

17   When a sequence has a constant ratio between successive terms it is called a **geometric progression** (often abbreviated to G.P.). The constant is called the **common ratio**, r. Examples include (i) 1, 2, 4, 8, ... where the common ratio is 2 and (ii) a, ar, ar^2, ar^3, ... where the common ratio is r.

18   If the first term of a G.P. is 'a' and the common ratio is r, then the n'th term is ar^{n-1}, which can be readily checked from the example in para 17.

For example, the 8'th term of the G.P. 1, 2, 4, 8, ... is $(1)(2)^7 = \mathbf{128}$, since $a=1$ and $r=2$.

Let a G.P. be a, ar, ar^2, ar^3, ... ar^{n-1}

then the sum of n terms, $S_n = a + ar + ar^2 + ar^3 + \ldots + ar^{n-1}$ (1)

Multiplying through by r gives:

$$rS_n = ar + ar^2 + ar^3 + ar^4 + \ldots ar^{n-1} + ar^n \ldots$$ (2)

Subtracting equation (2) from equation (1) gives

$S_n - rS_n = a - ar^n$; i.e. $S_n(1-r) = a(1-r^n)$

Thus the sum of n terms, $S_n = \dfrac{a(1-r^n)}{(1-r)}$, which is valid when $r < 1$.

Subtracting equation (1) from equation (2) gives

$$S_n = \frac{a(r^n - 1)}{(r-1)}$$

which is valid when $r > 1$.

For example, the sum of the first 8 terms of the GP 1, 2, 4, 8, 16, ... is given by

$$S_8 = \frac{1(2^8 - 1)}{(2-1)}, \text{ since } a = 1 \text{ and } r = 2,$$

i.e. $S_8 = \dfrac{1(256 - 1)}{1} = \mathbf{255}.$

19   When the common ratio r of a GP is less than unity, the sum of n terms,

$$S_n = \frac{a(1-r^n)}{(1-r)},$$

which may be written as

$$S_n = \frac{a}{(1-r)} - \frac{ar^n}{(1-r)}.$$

Since $r < 1$, r^n becomes less as n increases,

i.e. $r^n \to 0$ as $n \to \infty$.

Hence $\dfrac{ar^n}{(1-r)} \to 0$ as $n \to \infty$

Thus $S_n \to \dfrac{a}{(1-r)}$ as $n \to \infty$.

20   The quantity $\dfrac{a}{(1-r)}$ is called the **sum to infinity**, S_∞, and is the limiting value of the sum of an infinite number of terms,

i.e. $\boldsymbol{S_\infty = \dfrac{a}{1-r}}$ which is valid when $-1 < r < 1$.

For example, the sum to infinity of the GP $1 + \dfrac{1}{2} + \dfrac{1}{4} + \dots$ is

$$S_\infty = \cfrac{1}{1 - \cfrac{1}{2}},$$

since $a = 1$ and $r = \dfrac{1}{2}$, i.e., $S_\infty = 2$.

Fourier series
Periodic functions of period 2π

21   Fourier series provides a method of analysing periodic functions into their constituent components. Alternating currents and voltages, displacement, velocity and acceleration of slider-crank mechanisms and acoustic waves are typical practical examples in engineering and science where periodic functions are involved and often requiring analysis.

22   A function $f(x)$ is said to be **periodic** if $f(x + T) = f(x)$ for all values of x, where T is some positive number. T is the interval between two successive repetitions and is called the **period** of the functions $f(x)$. For example, $y = \sin x$ is periodic in x with period 2π since $\sin x = \sin(x + 2\pi) = \sin(x + 4\pi)$, and so on.

Similarly, $y = \cos x$ is a periodic function with period 2π since $\cos x = \cos(x + 2\pi) = \cos(x + 4\pi)$, and so on. In general, if $y = \sin \omega t$ or $y = \cos \omega t$ then the period of the waveform is $\dfrac{2\pi}{\omega}$.

The function shown in *Figure 4.1* is also periodic of period 2π and is defined by:

$$f(x) = \begin{cases} -1, & \text{when } -\pi < x < 0 \\ 1, & \text{when } 0 < x < \pi \end{cases}$$

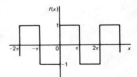

Figure 4.1

23   If a graph of a function has no sudden jumps or breaks it is called a **continuous function**, examples being the graphs of sine and cosine functions. However, other graphs make finite jumps at a point or points in the interval. The square wave shown in *Figure 4.1* has **finite discontinuities** at $x = \pi$, 2π, 3π, and so on. A great advantage of Fourier series over other series is that it can be applied to functions which are discontinuous as well as those which are continuous.

(i) The basis of a Fourier series is that all functions of practical significance which are defined in the interval $-\pi \leqslant x \leqslant \pi$ can be expressed in terms of a convergent **trigonometric series** of the form:

$$f(x) = a_0 + a_1 \cos x + a_2 \cos 2x + a_3 \cos 3x + \dots$$
$$+ b_1 \sin x + b_2 \sin 2x + b_3 \sin 3x + \dots,$$

when $a_0, a_1, a_2, \dots b_1, b_2, \dots$ are real constants,

i.e., $$f(x) = a_0 + \sum_{n=1}^{\infty} (a_n \cos nx + b_n \sin nx) \qquad (1)$$

where for the range $-\pi$ to π:

$$a_0 = \frac{1}{2\pi} \int_{-\pi}^{\pi} f(x) \, dx$$

$$a_n = \frac{1}{\pi} \int_{-\pi}^{\pi} f(x) \cos nx \, dx \quad (n = 1, 2, 3, \dots)$$

and $$b_n = \frac{1}{\pi} \int_{-\pi}^{\pi} f(x) \sin nx \, dx \quad (n = 1, 2, 3, \dots)$$

(ii) a_0, a_n and b_n are called the **Fourier coefficients** of the series and if these can be determined, the series of equation (1) is called the Fourier series corresponding to $f(x)$.

(iii) An alternative way of writing the series is by using the $a \cos x + b \sin x = c \sin(x + \alpha)$ relationship, i.e.,

$$f(x) = a_0 + c_1 \sin(x + \alpha_1) + c_2 \sin(2x + \alpha_2) + \dots$$
$$+ c_n \sin(nx + \alpha_n),$$

where a_0 is a constant, $c_1 = \sqrt{(a_1^2 + b_1^2)}, \dots c_n = \sqrt{(a_n^2 + b_n^2)}$ are the amplitudes of the various components, and phase angle $\alpha_n = \arctan\left(\dfrac{a_n}{b_n}\right)$.

(iv) For the series of equation (1):

the term $(a_1 \cos x + b_1 \sin x)$ or $c_1 \sin (x + \alpha_1)$ is called the **first harmonic** or the **fundamental**, the term $(a_2 \cos 2x + b_2 \sin 2x)$ or $(2x + \alpha_2)$ is called the **second harmonic**, and so on.

25   For an exact representation of a complex wave, an infinite number of terms are, in general, required. In many practical cases, however, it is sufficient to take the first few terms only.

26   The **sum of a Fourier series at a point of discontinuity** is given by the arithmetic mean of the two limiting values of $f(x)$ as x approaches the point of discontinuity from the two sides. For example, for the waveform shown in *Figure 4.2*, the sum of the Fourier series at the points of discontinuity $\left(\text{i.e. at } 0, \dfrac{\pi}{2}, \pi, \ldots \right)$ is given by:

$$\frac{8 + (-3)}{2} = \frac{5}{2} \text{ or } 2\frac{1}{2}.$$

A Fourier series for the square wave function shown in *Figure 4.3*, which is periodic of period 2π, is obtained as follows:

Since $f(x)$ is given by two different expressions in the two halves of the range the integration is done in two parts, one from $-\pi$ to 0 and the other from 0 to π.

From para 24(i)

$$a_0 = \frac{1}{2\pi} \int_{-\pi}^{\pi} f(x) \, dx = \frac{1}{2\pi} \left[\int_{-\pi}^{0} -k \, dx + \int_{-\pi}^{\pi} k \, dx \right]$$

$$= \frac{1}{2\pi} \{ [-kx]_{-\pi}^{0} + [kx]_{0}^{\pi} \} = 0$$

(a_0 is in fact the **mean value** of the waveform over a complete period of 2π and this could have been deduced in sight from *Figure 4.3*.)

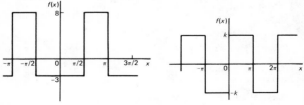

Figure 4.2   Figure 4.3

48

From para 24(i):

$$a_n = \frac{1}{\pi} \int\limits_{-\pi}^{\pi} f(x)\, \cos nx\, dx = \frac{1}{\pi} \left\{ \int\limits_{-\pi}^{0} - k\, \cos nx\, dx + \int\limits_{0}^{\pi} k\, \cos nx\, dx \right\}$$

$$= \frac{1}{\pi} \left\{ \left[\frac{-k\, \sin nx}{n} \right]_{-\pi}^{0} + \left[\frac{k\, \sin nx}{n} \right]_{0}^{\pi} \right\} = 0$$

Hence $a_1, a_2, a_3, \ldots$ are all zero and no cosine terms will appear in the Fourier series. From para 24(i):

$$b_n = \frac{1}{\pi} \int\limits_{-\pi}^{\pi} f(x)\, \sin nx\, dx = \frac{1}{\pi} \left\{ \int\limits_{-\pi}^{0} - k\, \sin nx\, dx + \int\limits_{0}^{\pi} k\, \sin nx\, dx \right\}$$

$$= \frac{1}{\pi} \left\{ \left[\frac{k\, \cos nx}{n} \right]_{-\pi}^{0} + \left[\frac{-k\, \cos nx}{n} \right]_{0}^{\pi} \right\}$$

When n is odd:

$$b_n = \frac{k}{\pi} \left\{ \left[\left(\frac{1}{n} \right) - \left(-\frac{1}{n} \right) \right] + \left[-\left(-\frac{1}{n} \right) - \left(-\frac{1}{n} \right) \right] \right\} = \frac{k}{\pi} \left\{ \frac{2}{n} + \frac{2}{n} \right\},$$

i.e. $b_n = \dfrac{4k}{n\pi}$

Hence

$$b_1 = \frac{4k}{\pi}, \; b_3 = \frac{4k}{3\pi}, \; b_5 = \frac{4k}{5\pi}, \text{ and so on.}$$

When n is even:

$$b_n = \frac{k}{\pi} \left\{ \left[\frac{1}{n} - \frac{1}{n} \right] + \left[-\frac{1}{n} - \left(-\frac{1}{n} \right) \right] \right\} = 0$$

Hence the Fourier series for the function in *Figure 4.3* is given by:

$$f(x) = \frac{4k}{\pi} \left(\sin x + \frac{1}{3} \sin 3x + \frac{1}{5} \sin 5x + \ldots \right)$$

If $k = \pi$ in this series then:

$$f(x) = 4 \left(\sin x + \frac{1}{3} \sin 3x + \frac{1}{5} \sin 5x + \ldots \right)$$

4 sin x is termed the first partial sum of the Fourier series of $f(x)$, $\left(4 \sin x + \dfrac{4}{3} \sin 3x\right)$ is termed the second partial sum of the Fourier series, and $\left(4 \sin x + \dfrac{4}{3} \sin 3x + \dfrac{4}{5} \sin 5x\right)$ is termed the third partial sum, and so on. Let $P_1 = 4 \sin x$, $P_2 = \left(4 \sin x + \dfrac{4}{3} \sin 3x\right)$ and $P_3 = \left(4 \sin x + \dfrac{4}{3} \sin 3x + \dfrac{4}{5} \sin 5x\right)$.

Graphs of P_1, P_2, and P_3, obtained by drawing up tables of values and adding waveforms, are shown in *Figure 4.4(a)* to *(c)* and they show that the series is convergent, i.e. continually approximating towards a definite limit as more and more partial sums are taken, and in the limit will have the sum $f(x) = \pi$.

Non-period functions over range 2π

27   If a function $f(x)$ is not periodic then it cannot be expanded in a Fourier series for **all** values of x. However, it is possible to determine a Fourier series to represent the function over any range of width 2π.

28   Given a non-periodic function, a new function may be constructed by taking the values of $f(x)$ in the given range and then repeating them outside of the given range at intervals of 2π. Since this new function is, by construction, periodic with period 2π, it may then be expanded in a Fourier series for all values of x. For example, the function $f(x) = x$ is not a periodic function.

However, if a Fourier series for $f(x) = x$ is required then the function is constructed outside of this range so that it is periodic with period 2π as shown by the broken lines in *Figure 4.5*.

29   For non-periodic functions, such as $f(x) = x$, the sum of the Fourier series is equal to $f(x)$ at all points in the given range but it is not equal to $f(x)$ at points outside of the range.

30   For determining a Fourier series of a non-periodic function over a range 2π, exactly the same formulae for the Fourier coefficients are used as in para 24(i).

A Fourier series to represent the function $f(x) = 2x$ in the range $-\pi$ to π is obtained as follows:

The function $f(x) = 2x$ is not periodic. The function is shown in the range $-\pi$ to π in *Figure 4.6* and is then constructed outside of that range so that it is periodic of period 2π (see broken lines) with the resulting saw-tooth waveform.

For a Fourier series:

$$f(x) = a_0 + \sum_{n=1}^{\infty} (a_n \cos nx + b_n \sin nx)$$

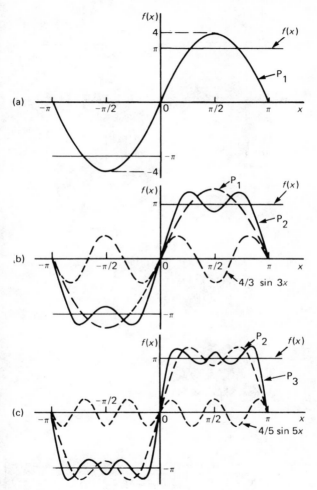

Figure 4.4

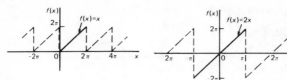

Figure 4.5

Figure 4.6

From para 24(i):

$$a_0 = \frac{1}{2\pi} \int\limits_{-\pi}^{\pi} f(x)\,dx = \frac{1}{2\pi} \int\limits_{-\pi}^{\pi} 2x\,dx = \frac{2}{2\pi}\left[\frac{x^2}{2}\right]_{-\pi}^{\pi} = 0$$

$$a_n = \frac{1}{\pi} \int\limits_{-\pi}^{\pi} f(x)\,\cos\,nx\,dx = \frac{1}{\pi} \int\limits_{-\pi}^{\pi} 2x\,\cos\,nx\,dx$$

$$= \frac{2}{\pi}\left[\frac{x\,\sin\,x}{n} - \int\frac{\sin\,nx}{n}\,dx\right]_{-\pi}^{\pi},$$

by parts (see paras 13 and 14, page 194)

$$= \frac{2}{\pi}\left[\frac{x\,\sin\,nx}{n} + \frac{\cos\,nx}{n^2}\right]_{-\pi}^{\pi}$$

$$= \frac{2}{\pi}\left[\left(0 + \frac{\cos\,n\pi}{n^2}\right) - \left(0 + \frac{\cos\,n\,(-\pi)}{n^2}\right)\right] = 0$$

$$b_n = \frac{1}{\pi} \int\limits_{-\pi}^{\pi} f(x)\,\sin\,nx\,dx = \frac{1}{\pi} \int\limits_{-\pi}^{\pi} 2x\,\sin\,nx\,dx$$

$$= \frac{2}{\pi}\left[\frac{-x\,\cos\,nx}{n} - \int\left(\frac{-\cos\,nx}{n}\right)dx\right]_{-\pi}^{\pi},\text{ by parts}$$

$$= \frac{2}{\pi}\left[\frac{-x\,\cos\,nx}{n} + \frac{\sin\,nx}{n^2}\right]_{-\pi}^{\pi}$$

$$= \frac{2}{\pi}\left[\left(\frac{-\pi\,\cos\,n\pi}{n} + \frac{\sin\,n\pi}{n^2}\right) - \left(\frac{-(-\pi)\,\cos\,n\,(-\pi)}{n} + \frac{\sin\,n\,(-\pi)}{n^2}\right)\right]$$

$$= \frac{2}{\pi}\left[\frac{-\pi\,\cos\,n\pi}{n} - \frac{\pi\,\cos\,n\pi}{n}\right] = \frac{-4}{n}\,\cos\,n\pi$$

52

When n is odd:

$$b_n = \frac{4}{n}.$$

Thus $b_1 = 4$, $b_3 = \frac{4}{3}$, $b_5 = \frac{4}{5}$, and so on.

When n is even:

$$b_n = \frac{-4}{n}.$$

Thus $b_2 = -2$, $b_4 = -1$, $b_6 = -\frac{2}{3}$, and so on.

Thus $f(x) = 2x = 4 \sin x - 2 \sin 2x + \frac{4}{3} \sin 3x - 1 \sin 4x + \frac{4}{5} \sin 5x$

$$- \frac{2}{3} \sin 6x + \dots$$

i.e. $\mathbf{2x = 4\left(\sin x - \frac{1}{2}\sin 2x + \frac{1}{3}\sin 3x - \frac{1}{4}\sin 4x + \frac{1}{5}\sin 5x - \frac{1}{6}\sin 6x + \dots \right)}$

for values of $f(x)$ between $-\pi$ and π. For values of $f(x)$ outside the range $-\pi$ to π the sum of the series is not equal to $f(x)$.

Even and odd functions

Even functions

31   A function $y = f(x)$ is said to be even if $f(-x) = f(x)$ for all values of x. Graphs of even functions are always **symmetrical about the y-axis** (i.e. is a mirror image). Two examples of even functions are $y = x^2$ and $y = \cos x$ as shown in *Figure 4.7*. An even function contains no sine terms in its Fourier series, i.e., $\boldsymbol{b_n = 0}$

Odd functions

32   A function $y = f(x)$ is said to be odd if $f(-x) = -f(x)$ for all values of x. Graphs of odd functions are always **symmetrical about the origin.** Two examples of odd functions are $y = x^5$ and $y = \sin x$ as shown in *Figure 4.8*. An odd function contains no cosine terms in its Fourier series, i.e., $\boldsymbol{a_n = 0}$ In addition, the constant term $\boldsymbol{a_0 = 0}$.

33   Many functions are neither even nor odd, two such examples being shown in *Figure 4.9*.

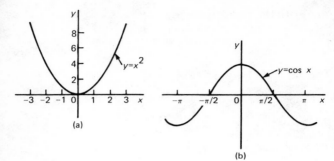

Figure 4.7

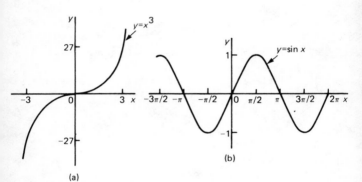

Figure 4.8

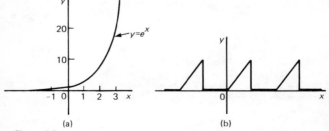

Figure 4.9

54

Fourier cosine series

34 The Fourier series of an even periodic function $f(x)$ having period 2π contains **cosine terms only** (i.e. contains no sine terms) and may contain a constant term. Hence,

$$f(x) = a_0 + \sum_{n=1}^{\infty} a_n \cos nx$$

where

$$a_0 = \frac{1}{2\pi} \int_{-\pi}^{\pi} f(x)\ dx = \frac{1}{\pi} \int_{0}^{\pi} f(x)\ dx \quad \text{(due to symmetry)}$$

and

$$a_n = \frac{1}{\pi} \int_{-\pi}^{\pi} f(x)\cos nx\ dx = \frac{2}{\pi} \int_{0}^{\pi} f(x)\cos nx\ dx$$

The square wave shown in *Figure 4.10* is an even function since it is symmetrical about the $f(x)$ axis.

Hence the Fourier series is given by

$$f(x) = a_0 + \sum_{n=1}^{\infty} \cos nx$$

(i.e. the series contains no sine terms).

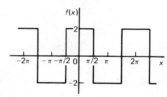

Figure 4.10

$a_0 = 0$ (i.e. the mean value)

$$a_n = \frac{2}{\pi} \int_{0}^{\pi} f(x)\cos nx\ dx = \frac{2}{\pi}\left\{ \int_{0}^{\pi/2} 2\cos nx\ dx + \int_{\pi/2}^{\pi} -2\cos nx\ dx \right\}$$

$$= \frac{4}{\pi}\left\{ \left[\frac{\sin nx}{n}\right]_{0}^{\pi/2} + \left[\frac{-\sin nx}{n}\right]_{\pi/2}^{\pi} \right\} = \frac{4}{\pi}\left\{ \left(\frac{\sin \frac{\pi}{2}n}{n} - 0\right) + \left(0 - \frac{-\sin \frac{\pi}{2}n}{n}\right) \right\}$$

$$= \frac{4}{\pi}\left(\frac{2\sin \frac{\pi}{2}n}{n}\right) = \frac{8}{\pi n}\left(\sin \frac{n\pi}{2}\right)$$

When n is even;

$$a_n = 0$$

55

When n is odd;

$$a_n = \frac{8}{\pi n} \text{ for } n = 1, 5, 9, \dots \text{ and } a_n = \frac{-8}{\pi n} \text{ for } n = 3, 7, 11, \dots$$

Hence

$$a_1 = \frac{8}{\pi}, \; a_3 = \frac{-8}{3\pi}, \; a_5 = \frac{8}{5\pi}, \text{ and so on.}$$

Hence the Fourier series for the waveform of *Figure 4.10* is given by:

$$f(x) = \frac{8}{\pi}\left(\cos x - \frac{1}{3} \cos 3x + \frac{1}{5} \cos 5x - \frac{1}{7} \cos 7x + \dots \right)$$

Fourier sine series

35   The Fourier series of an odd periodic function $f(x)$ having period 2π contains **sine terms only** (i.e., contains no constant term and no cosine terms).

Hence, $f(x) = \sum\limits_{n=1}^{\infty} b_n \, \textbf{sin } nx$

where $b_n = \dfrac{1}{\pi} \int\limits_{-\pi}^{\pi} f(x) \sin nx \; dx = \dfrac{2}{\pi} \int\limits_{0}^{\pi} f(x) \sin nx \; dx$

The square wave shown in *Figure 4.11* is an odd function since it is symmetrical about the origin.

Hence the Fourier series is given by:

$$f(x) = \sum\limits_{n=1}^{\infty} b_n \sin nx$$

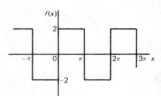

The function is given by:

$$f(x) = \begin{cases} -2, & \text{when } -\pi < x < 0 \\ 2, & \text{when } 0 < x < \pi \end{cases}$$

Figure 4.11

$$b_n = \frac{2}{\pi} \int\limits_{0}^{\pi} f(x) \sin nx \; dx = \frac{2}{\pi} \int\limits_{0}^{\pi} 2 \sin nx \; dx = \frac{4}{\pi} \left[\frac{-\cos nx}{n} \right]_{0}^{\pi}$$

$$= \frac{4}{\pi} \left[\left(\frac{-\cos n\pi}{n} \right) - \left(-\frac{1}{n} \right) \right] = \frac{4}{\pi n} (1 - \cos n\pi).$$

When n is even $b_n = 0$.

When n is odd $b_n = \dfrac{4}{\pi n}(1 - -1) = \dfrac{8}{\pi n}$.

Hence $b_1 = \dfrac{8}{\pi}$, $b_3 = \dfrac{8}{3\pi}$, $b_5 = \dfrac{8}{5\pi}$, and so on.

Hence the Fourier series for the square waveform shown in *Figure 4.11* is:

$$f(x) = \frac{8}{\pi}\left(\sin x + \frac{1}{3} \sin 3x + \frac{1}{5} \sin 5x + \frac{1}{7} \sin 7x + \ldots \right)$$

Half range Fourier series

36   When a function is defined over the range say 0 to π instead of from 0 to 2π it may be expanded in a series of sine terms only or of cosine terms only. The series produced is called a half range Fourier series.

37   If a **half range cosine series** is required for the function $f(x) = x$ in the range 0 to π then an **even** periodic function is required. In *Figure 4.12*, $f(x) = x$ is shown plotted from $x = 0$ to $x = \pi$. Since an even function is symmetrical about the $f(x)$ axis the line AB is constructed as shown. If the triangular waveform produced is assumed to be periodic of period 2π outside of this range then the waveform is as shown in *Figure 4.12*. When a half range cosine series is required then the Fourier coefficients a_0 and a_n are calculated as in para 34.

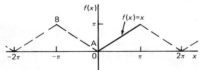

Figure 4.12

Hence for a half range cosine series:

$$f(x) = a_0 + \sum_{n=1}^{\pi} a_n \cos nx.$$

When

$$f(x) = x, \; a_0 = \frac{1}{\pi}\int\limits_{0}^{\pi} f(x) \; dx = \frac{1}{\pi}\int\limits_{0}^{\pi} x \; dx = \frac{1}{\pi}\left[\frac{x^2}{2} \right]_0^{\pi} = \frac{\pi}{2}$$

57

$$a_n = \frac{2}{\pi} \int\limits_0^\pi f(x) \cos nx \; dx = \frac{2}{\pi} \int\limits_0^\pi x \cos nx \; dx = \frac{2}{\pi} \left[\frac{x \sin nx}{n} + \frac{\cos nx}{n^2} \right]_0^\pi \text{ by parts}$$

$$= \frac{2}{\pi} \left[\left(\frac{\pi \sin n\pi}{n} + \frac{\cos n\pi}{n^2} \right) - \left(0 + \frac{\cos 0}{n^2} \right) \right]$$

$$= \frac{2}{\pi} \left(0 + \frac{\cos n\pi}{n^2} - \frac{\cos 0}{n^2} \right) = \frac{2}{\pi n^2} (\cos n\pi - 1)$$

When n is even

$$a_n = 0.$$

When n is odd

$$a_n = \frac{2}{\pi n^2} (-1-1) = \frac{-4}{\pi n^2}$$

Hence $a_1 = \frac{-4}{\pi}$, $a_3 = \frac{-4}{\pi 3^2}$, $a_5 = \frac{-4}{\pi 5^2}$, and so on.

Hence the half range Fourier cosine series is given by:

$$f(x) = x = \frac{\pi}{2} - \frac{4}{\pi} \left(\cos x + \frac{1}{3^2} \cos 3x + \frac{1}{5^2} \cos 5x + \ldots \right)$$

38   If a **half range sine series** is required for the function $f(x) = x$ in the range 0 to π then an odd periodic function is required. In *Figure 4.13*, $f(x) = x$ is shown plotted from $x = 0$ to $x = \pi$.

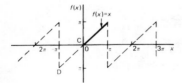

Figure 4.13

Since an odd function is symmetrical about the origin the line CD is constructed as shown. If the sawtooth waveform produced is assumed to be periodic of period 2π outside of this range, then the waveform is as shown in *Figure 4.13*. When a half range sine series is require then the Fourier coefficient b_n is calculated as in para 35.

Hence for a half range sine series: $f(x) = \sum\limits_{n=1}^{\infty} b_n \sin nx$.

When $f(x) = x$,

$$b_n = \frac{2}{\pi} \int\limits_0^{\pi} f(x) \sin nx \, dx = \frac{2}{\pi} \int\limits_0^{\pi} x \sin nx \, dx$$

$$= \frac{2}{\pi} \left[\frac{-x \cos nx}{n} + \frac{\sin nx}{n^2} \right]_0^{\pi}, \text{ by parts,}$$

$$= \frac{2}{\pi} \left[\left(\frac{-\pi \cos n\pi}{n} + \frac{\sin n\pi}{n^2} \right) - (0+0) \right] = -\frac{2}{n} \cos n\pi$$

When n is odd,

$$b_n = \frac{2}{n}.$$

Hence $b_1 = \frac{2}{1}$, $b_3 = \frac{2}{3}$, $b_5 = \frac{2}{5}$, and so on.

When n is even

$$b_n = -\frac{2}{n}.$$

Hence $b_2 = -\frac{2}{2}$, $b_4 = -\frac{2}{4}$, $b_6 = -\frac{2}{6}$, and so on.

Hence the half range Fourier sine series is given by:

$$f(x) = x = 2\left(\sin x - \frac{1}{2} \sin 2x + \frac{1}{3} \sin 3x - \frac{1}{4} \sin 4x + \frac{1}{5} \sin 5x - \dots \right)$$

Fourier series over any range

39   A periodic function $f(x)$ of period l repeats itself when x increases by l, i.e. $f(x+l) = f(x)$. The change from functions dealt with previously having period 2π to functions having period l is not difficult since it may be achieved by a change of variable.

40   To find a Fourier series for a function $f(x)$ in the range $-\frac{1}{2} \leqslant x \leqslant \frac{l}{2}$ a new variable u is introduced such that $f(x)$, as a function of u, has period 2π. If $u = \frac{2\pi x}{l}$ then, when $x = -\frac{l}{2}$, $u = -\pi$ and when $x = \frac{l}{2}$, $u = +\pi$. Also, let $f(x) = f\left(\frac{lu}{2\pi}\right) = F(u)$. The

Fourier series for $F(u)$ is given by:

$$F(u) = a_0 + \sum_{n=1}^{\infty} (a_n \cos nu + b_n \sin nu)$$

where $a_0 = \dfrac{1}{2\pi} \displaystyle\int_{-\pi}^{\pi} F(u) \, du, \quad a_n = \dfrac{1}{\pi} \displaystyle\int_{-\pi}^{\pi} F(u) \cos nu \, du$ and

$$b_n = \frac{1}{\pi} \int_{-\pi}^{\pi} F(u) \sin nu \, du.$$

41 It is however more usual to change the formula of para 40 to terms of x. Since $u = \dfrac{2\pi x}{l}$, then $du = \dfrac{2\pi}{l} \, dx$, and the limits of integration are $-\dfrac{l}{2}$ to $+\dfrac{l}{2}$ instead of from $-\pi$ to $+\pi$. Hence the Fourier series expressed in terms of x is given by:

$$f(x) = a_0 + \sum_{n=1}^{\infty} \left\{ a_n \cos\left(\frac{2\pi n x}{l}\right) + b_n \sin\left(\frac{2\pi n x}{l}\right) \right\},$$

where, in the range $-\dfrac{l}{2}$ to $+\dfrac{l}{2}$:

$$a_0 = \frac{1}{l} \int_{-l/2}^{l/2} f(x) \, dx, \quad a_n = \frac{2}{l} \int_{-l/2}^{l/2} f(x) \cos\left(\frac{2\pi n x}{l}\right) dx$$

$$\text{and } b_n = \frac{2}{l} \int_{-l/2}^{l/2} f(x) \sin\left(\frac{2\pi n x}{l}\right) dx$$

(The limits of integration may be replaced by any interval of length l, such as from 0 to l.)

For example, if the voltage from a square wave generator is of the form:

$$v(t) = \begin{cases} 0, & -4 < t < 0 \\ 10, & 0 < t < 4 \end{cases}$$

and has a period of 8 ms, then the Fourier series is obtained as follows. The square wave is shown in *Figure 4.14*.

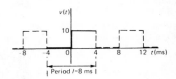

Figure 4.14

The Fourier series is of the form:

$$v(t) = a_0 + \sum_{n=1}^{\infty}\left[a_n \cos\left(\frac{2\pi\, nt}{l}\right) + b_n \sin\left(\frac{2\pi nt}{l}\right)\right]$$

$$a_0 = \frac{1}{l}\int_{-l/2}^{l/2} v(t)\ dt = \frac{1}{8}\int_{-4}^{4} v(t)\ dt = \frac{1}{8}\left\{\int_{-4}^{0} 0\ dt + \int_{0}^{4} 10\ dt\right\} = \frac{1}{8}[10t]_0^4 = 5$$

$$a_n = \frac{2}{l}\int_{-l/2}^{l/2} v(t)\ \cos\left(\frac{2\pi nt}{l}\right)dt = \frac{2}{8}\int_{-4}^{4} v(t)\ \cos\left(\frac{2\pi nt}{8}\right)\ dt$$

$$= \frac{1}{4}\left\{\int_{-4}^{0} 0\ \cos\left(\frac{\pi nt}{4}\right)dt + \int_{0}^{4} 10\ \cos\left(\frac{\pi nt}{4}\right)dt\right\}$$

$$= \frac{1}{4}\left[\frac{10 \sin\left(\dfrac{\pi nt}{4}\right)}{\left(\dfrac{\pi n}{4}\right)}\right]_0^4 = \frac{10}{\pi n}[\sin\,\pi n - \sin\,0] = 0$$

for $n = 1, 2, 3, \ldots$

$$b_n = \frac{2}{l}\int_{-l/2}^{l/2} v(t)\ \sin\left(\frac{2\pi nt}{l}\right)\ dt = \frac{2}{8}\int_{-4}^{4} v(t)\ \sin\left(\frac{2\pi nt}{8}\right)\ dt$$

$$= \frac{1}{4}\left\{\int_{-4}^{0} 0\ \sin\left(\frac{\pi nt}{4}\right)\ dt + \int_{0}^{4} 10\ \sin\left(\frac{\pi nt}{4}\right)\ dt\right\}$$

$$= \frac{1}{4}\left[\frac{-10 \cos\dfrac{\pi nt}{4}}{\dfrac{\pi n}{4}}\right]_0^4 = \frac{-10}{\pi n}[\cos\,\pi n - \cos\,0]$$

When n is even

$$b_n = 0.$$

When n is odd,

$$b_1 = \frac{-10}{\pi}(-1-1) = \frac{20}{\pi},\ b_3 = \frac{-10}{3\pi}(-1-1) = \frac{20}{3\pi},\ b_5 = \frac{20}{5\pi}\ \text{and so on.}$$

Thus the Fourier series for the function $v(t)$ is given by:

$$v(t) = 5 + \frac{20}{\pi}\left[\sin\frac{\pi t}{4} + \frac{1}{3}\sin\left(\frac{3\pi t}{4}\right) + \frac{1}{5}\sin\left(\frac{5\pi t}{4}\right) + \dots\right]$$

Half range series

42   By making the substitution $u = \frac{\pi x}{l}$ (see para 40), the range $x = 0$ to $x = l$ corresponds to the range $u = 0$ to $u = \pi$. Hence a function may be expanded in a series of either cosine terms or sine terms only, i.e. a **half range Fourier series**.

43   A **half range cosine series** in the range 0 to l can be expanded as:

$$f(x) = a_0 + \sum_{n=1}^{\infty} a_n \cos\left(\frac{n\pi x}{l}\right)$$

where $a_0 = \frac{1}{l}\int_0^l f(x)\,dx$ **and** $a_n = \frac{2}{l}\int_0^l f(x)\cos\left(\frac{n\pi x}{l}\right)\,dx$

For example, the half range Fourier cosine series for the function $f(x) = x$, in the range $0 \leqslant x \leqslant 2$ is obtained as follows: A half range Fourier cosine series indicates an even function. Thus the graph of $f(x) = x$ in the range 0 to 2 is shown in *Figure 4.15* and is extended outside of this range so as to be symmetrical about the $f(x)$ axis as shown by the broken lines.

For a half range cosine series:

$$f(x) = a_0 + \sum_{n=1}^{\infty} a_n \cos\left(\frac{n\pi x}{l}\right)$$

$$a_0 = \frac{1}{l}\int_0^l f(x)\,dx = \frac{1}{2}\int_0^2 x\,dx = \frac{1}{2}\left[\frac{x^2}{2}\right]_0^2 = 1$$

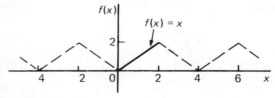

Figure 4.15

62

$$a_n = \frac{2}{l}\int\limits_0^l f(x)\,\cos\left(\frac{n\pi x}{l}\right)dx = \frac{2}{2}\int\limits_0^2 x\,\cos\left(\frac{n\pi x}{2}\right)dx = \left[\frac{x\,\sin\left(\dfrac{n\pi x}{2}\right)}{\left(\dfrac{n\pi}{2}\right)} + \frac{\cos\left(\dfrac{n\pi x}{2}\right)}{\left(\dfrac{n\pi}{2}\right)^2}\right]_0^2$$

$$= \left[\left(\frac{2\,\sin\,n\pi}{\left(\dfrac{n\pi}{2}\right)} + \frac{\cos\,n\pi}{\left(\dfrac{n\pi}{2}\right)^2}\right) - \left(0 + \frac{\cos\,0}{\left(\dfrac{n\pi}{2}\right)^2}\right)\right]$$

$$= \left[\frac{\cos\,n\pi}{\left(\dfrac{n\pi}{2}\right)^2} - \frac{1}{\left(\dfrac{n\pi}{2}\right)^2}\right] = \left(\frac{2}{\pi n}\right)^2(\cos\,n\pi - 1)$$

When n is even

$a_n = 0$.

$b_1 = \dfrac{-8}{\pi^2}$, $b_3 = \dfrac{-8}{\pi^2 3^2}$, $b_5 = \dfrac{-8}{\pi^2 5^2}$, and so on.

Hence the half range Fourier cosine series for $f(x)$ in the range 0 to 2 is given by:

$$f(x) = 1 - \frac{8}{\pi^2}\left[\cos\left(\frac{\pi x}{2}\right) + \frac{1}{3^2}\cos\left(\frac{3\pi x}{2}\right) + \frac{1}{5^2}\cos\left(\frac{5\pi x}{2}\right) + \dots\right]$$

44 A **half range sine series** in the range 0 to l can be expanded as:

$$f(x) = \sum_{n=1}^{\infty} b_n\,\sin\left(\frac{n\pi x}{l}\right),$$

$$\text{where } b_n = \frac{2}{l}\int\limits_0^l f(x)\,\sin\left(\frac{n\pi x}{l}\right)dx$$

For example, the half range Fourier sine series for the function $f(x) = x$ in the range $0 \leqslant x \leqslant 2$ is obtained as follows:

A half range Fourier sine series indicates an odd function. Thus the graphs of $f(x) = x$ in the range 0 to 2 is shown in *Figure 4.16* and is extended outside of this range so as to be symmetrical about the origin, as shown by the broken lines.

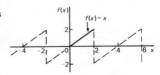

Figure 4.16

For a half range sine series:

$$f(x) = \sum_{n=1}^{\infty} b_n \sin\left(\frac{n\pi x}{l}\right)$$

$$b_n = \frac{2}{l} \int_0^l f(x) \sin\left(\frac{n\pi x}{l}\right) dx = \frac{2}{2} \int_0^2 x \sin\left(\frac{n\pi x}{2}\right) dx$$

$$= \left[\frac{-x \cos\left(\frac{n\pi x}{2}\right)}{\left(\frac{n\pi}{2}\right)} - \frac{\sin\left(\frac{n\pi x}{2}\right)}{\left(\frac{n\pi}{2}\right)^2} \right]_0^2$$

$$= \left[\left(\frac{-2 \cos n\pi}{\left(\frac{n\pi}{2}\right)} - \frac{\sin n\pi}{\left(\frac{n\pi}{2}\right)^2} \right) - \left(0 - \frac{\sin 0}{\left(\frac{n\pi}{2}\right)^2} \right) \right]$$

$$= \frac{-2 \cos n\pi}{\frac{n\pi}{2}} = \frac{-4}{n\pi} \cos n\pi$$

Hence $b_1 = \frac{-4}{\pi}(-1) = \frac{4}{\pi}$, $b_2 = \frac{-4}{2\pi}(1) = \frac{-4}{2\pi}$, $b_3 = \frac{-4}{3\pi}(1) = \frac{4}{3\pi}$, and so on.

Thus the half range Fourier sine series in the range 0 to 2 is given by:

$$f(x) = \frac{4}{\pi}\left[\sin\left(\frac{\pi x}{2}\right) - \frac{1}{2} \sin\left(\frac{2\pi x}{2}\right) + \frac{1}{3} \sin\left(\frac{3\pi x}{2}\right) - \frac{1}{4} \sin\left(\frac{4\pi x}{2}\right) + \ldots \right]$$

A numerical method of harmonic analysis

45   Many practical waveforms can be represented by simple mathematical expressions, and, by using Fourier series, the magnitude of their harmonic components determined. For waveforms not in this category, analysis may be achieved by numerical methods.

46   **Harmonic analysis** is the process of resolving a periodic, non-sinusoidal quantity into a series of sinusoidal components of ascending order of frequency.

47   The Fourier coefficients, a_0, a_n and b_n stated in para 24 all require functions to be integrated, i.e.

$$a_0 = \frac{1}{2\pi} \int_{-\pi}^{\pi} f(x) \; dx = \frac{1}{2\pi} \int_{0}^{2\pi} f(x) \; dx = \text{mean value of } f(x) \text{ in the}$$

$$\text{range } -\pi \text{ to } \pi \text{ or } 0 \text{ to } 2\pi.$$

$$a_n = \frac{1}{\pi} \int_{-\pi}^{\pi} f(x) \; \cos nx \; dx = \frac{1}{\pi} \int_{0}^{2\pi} f(x) \; \cos nx \; dx = \text{twice the mean}$$

$$\text{value of } f(x) \; \cos nx \text{ in the range } 0 \text{ to } 2\pi.$$

$$b_n = \frac{1}{\pi} \int_{-\pi}^{\pi} f(x) \; \sin nx \; dx = \frac{1}{\pi} \int_{0}^{2\pi} f(x) \; \sin nx \; dx = \text{twice the mean}$$

$$\text{value of } f(x) \; \sin nx \text{ in the range } 0 \text{ to } 2\pi.$$

However, irregular waveforms are not usually defined by mathematical expressions and thus Fourier coefficients cannot be determined by using calculus. In these cases, approximate methods, such as the **trapezoidal rule** can be used to evaluate the Fourier coefficients.

48   Most practical waveforms to be analysed are periodic. Let the period of a waveform be 2π and be divided into p equal parts as shown in *Figure 4.17*. The width of each interval is thus $\frac{2\pi}{p}$. Let the ordinates be labelled $y_0, y_1, y_2, \ldots y_p$, (note that $y_0 = y_p$). The trapezoidal rule states:

$$\text{Area} = \left(\begin{array}{c} \text{width of} \\ \text{interval} \end{array} \right) \left[\frac{1}{2} \left(\begin{array}{c} \text{first + last} \\ \text{ordinate} \end{array} \right) + \begin{array}{c} \text{sum of} \\ \text{remaining} \\ \text{ordinates} \end{array} \right]$$

$$\approx \left(\frac{2\pi}{p} \right) \left[\frac{1}{2}(y_0 + y_p) + y_1 + y_2 + y_3 + \ldots \right]$$

Since $y_0 = y_p$, then $\frac{1}{2}(y_0 + y_p) = y_0 = y_p$.

Hence area $\approx \dfrac{2\pi}{p} \displaystyle\sum_{k=1}^{p} y_k$

$$\text{Mean value} = \frac{\text{area}}{\text{length of base}} \approx \frac{1}{2\pi} \left(\frac{2\pi}{p} \right) \sum_{k=1}^{p} y_k \simeq \frac{1}{p} \sum_{k=1}^{p} y_k$$

However, $a_0 = $ mean value of $f(x)$ in the range 0 to 2π.

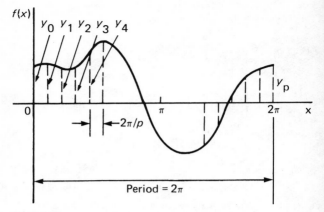

Figure 4.17

thus $a_0 \simeq \dfrac{1}{p} \sum\limits_{k=1}^{p} y_k$ (1)

Similarly, a_n = twice the mean value of $f(x) \cos nx$ in the range 0 to 2π,

thus $a_n \approx \dfrac{2}{p} \sum\limits_{k=1}^{p} y_k \cos nx_k$ (2)

and b_n = twice the mean value of $f(x) \sin nx$ in the range 0 to 2π,

thus $\left| b_n \approx \dfrac{2}{p} \sum\limits_{k=1}^{p} y_k \sin nx_k \right.$ (3)

A graph of voltage v against θ degrees is shown in *Figure 4.18*. The values of ordinates $y_1, y_2, y_3, \ldots$ are 62, 35, -38, -64, -63, -52, -28, 24, 80, 96, 90 and 70, the 12 equal intervals each being of width 30°. The voltage may be analysed into its first 3 constituent harmonics as follows:

The data is tabulated in the proforma shown in *Table 4.2*.

From equation (1), $a_0 \approx \dfrac{1}{p} \sum\limits_{k=1}^{p} y_k = \dfrac{1}{12}(212) = 17.67$ (since $p = 12$)

From equation (2), $a_n \approx \dfrac{2}{p} \sum\limits_{k=1}^{p} y_k \cos nx_k$

66

Table 4.2

Ordinates	$\theta°$	V	$\cos\theta$	$V\cos\theta$	$\sin\theta$	$V\sin\theta$	$\cos2\theta$	$V\cos2\theta$	$\sin2\theta$	$V\sin2\theta$	$\cos3\theta$	$V\cos3\theta$	$\sin3\theta$	$V\sin3\theta$
y_1	30	62	0.866	53.69	0.5	31	0.5	31	0.866	53.69	0	0	1	62
y_2	60	35	0.5	17.5	0.866	30.31	-0.5	-17.5	0.866	30.31	-1	-35	0	0
y_3	90	-38	0	0	1	-38	-1	38	0	0	0	0	-1	38
y_4	120	-64	-0.5	32	0.866	-55.42	-0.5	32	-0.866	55.42	1	-64	0	0
y_5	150	-63	-0.866	54.56	0.5	-31.5	0.5	-31.5	-0.866	54.56	0	0	1	-63
y_6	180	-52	-1	52	0	0	1	-52	0	0	-1	52	0	0
y_7	210	-28	-0.866	24.25	-0.5	14	0.5	-14	0.866	-24.25	0	0	-1	28
y_8	240	24	-0.5	-12	-0.866	-20.78	-0.5	-12	0.866	20.78	1	24	0	0
y_9	270	80	0	0	-1	-80	-1	-80	0	0	0	0	1	80
y_{10}	300	96	0.5	48	-0.866	-83.14	-0.5	-48	-0.866	-83.14	-1	-96	0	0
y_{11}	330	90	0.866	77.94	-0.5	-45	0.5	45	-0.866	-77.94	0	0	-1	-90
y_{12}	360	70	1	70	0	0	1	70	0	0	1	70	0	0
$\sum_{k=1}^{12} y_k$		$=212$		$\sum_{k=1}^{12} y_k \cos\theta_k$ $=417.94$		$\sum_{k=1}^{12} y_k \sin\theta_k$ $=-278.53$		$\sum_{k=1}^{12} y_k \cos2\theta_k$ $=-39$		$\sum_{k=1}^{12} y_k \sin2\theta_k$ $=29.43$		$\sum_{k=1}^{12} y_k \cos3\theta_k$ $=-49$		$\sum_{k=1}^{12} y_k \sin3\theta_k$ $=55$

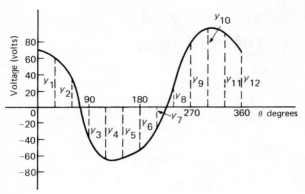

Figure 4.18

Hence $a_1 \approx \dfrac{2}{12}(417.94) = 69.66$; $a_2 \approx \dfrac{2}{12}(-39) = -6.50$;

and $a_3 \approx \dfrac{2}{12}(-49) = -8.17$

From equation (3), $b_n \approx \dfrac{2}{p} \displaystyle\sum_{k=1}^{p} y_k \sin n x_k$

Hence, $b_1 \approx \dfrac{2}{12}(-278.53) = -46.42$; $b_2 \approx \dfrac{2}{12}(29.43) = 4.91$;

and $b_3 \approx \dfrac{2}{12}(55) = 9.17$

Substituting these values into the Fourier series:

$$f(x) = a_0 + \sum_{n=1}^{\infty} (a_n \cos nx + b_n \sin nx)$$

gives

$$v = 17.67 + 69.66 \cos \theta - 6.50 \cos 2\theta - 8.17 \cos 3\theta + \ldots$$
$$- 46.42 \sin \theta + 4.91 \sin 2\theta + 9.17 \sin 3\theta + \ldots$$

5 Matrices and determinants

1   Matrices and determinants may be used for the solution of linear simultaneous equations. The coefficients of the variables for linear simultaneous equations may be shown in matrix form. The coefficients of x and y in the simultaneous equations

$$\begin{aligned} x + 2y &= 3 \\ 4x - 5y &= 6 \end{aligned} \quad \text{become} \quad \begin{pmatrix} 1 & 2 \\ 4 & -5 \end{pmatrix} \quad \text{in matrix notation}$$

Similarly, the coefficients of p, q and r in the equations

$$\begin{aligned} 1.3p - 2.0q + r &= 7 \\ 3.7p + 4.8q - 7r &= 3 \\ -4.1p + 3.8q + 12r &= -6 \end{aligned} \quad \text{become} \quad \begin{pmatrix} 1.3 & -2.0 & 1 \\ 3.7 & 4.8 & -7 \\ -4.1 & 3.8 & 12 \end{pmatrix}$$

in matrix form

2   The numbers within a matrix are called an **array** and the coefficients forming the array are called the **elements** of the matrix. The number of rows in a matrix is usually specified by m and the number of columns by n and a matrix referred to as an 'm by n' matrix. Thus,

$$\begin{pmatrix} 2 & 3 & 6 \\ 4 & 5 & 7 \end{pmatrix}$$

is a '2 by 3' matrix.

3   Matrices cannot be expressed as a single numerical value, but they can often be simplified or combined, and unknown element values can be determined by comparison methods. Just as there are rules for addition, subtraction, multiplication and division of numbers in arithmetic, rules for these operations can be applied to matrices and the rules of matrices are such that they obey most of those governing the algebra of numbers.

Addition and subtraction of matrices

4   If

$$A = \begin{pmatrix} a & b \\ c & d \end{pmatrix} \text{ and } B = \begin{pmatrix} e & f \\ g & h \end{pmatrix}$$

then

$$A + B = \begin{pmatrix} a+e & b+f \\ c+g & d+h \end{pmatrix}$$

and

$$A - B = \begin{pmatrix} a-e & b-f \\ c-g & d-h \end{pmatrix}$$

Multiplication of matrices

5   (i) When a matrix is multiplied by a number (called scalar multiplication), a single matrix results in which each element of the original matrix has been multiplied by the number. For example

$$3 \begin{pmatrix} 2 & -1 \\ 0 & 4 \end{pmatrix} = \begin{pmatrix} 6 & -3 \\ 0 & 12 \end{pmatrix}$$

(ii) When a matrix A is multiplied by another matrix B, a single matrix results in which elements are obtained from the sum of the products of the corresponding rows of A and the corresponding columns of B. For example,

$$\begin{pmatrix} a & b \\ c & d \end{pmatrix} \times \begin{pmatrix} e & f \\ g & h \end{pmatrix} = \begin{pmatrix} ae+bg & af+bh \\ ce+dg & cf+dh \end{pmatrix}$$

(iii) Two matrices A and B may be multiplied together, provided the number of elements in the rows of matrix A are equal to the number of elements in the columns of matrix B. In general terms, when multiplying a matrix of dimensions (m by n) by a matrix of dimensions (n by r), the resulting matrix has dimensions (m by r). Thus a 2 by 3 matrix multiplied by a 3 by 1 matrix gives a matrix of dimensions 2 by 1. Thus

$$\begin{pmatrix} 3 & 4 & 0 \\ -2 & 6 & -3 \\ 7 & -4 & 1 \end{pmatrix} \times \begin{pmatrix} 2 \\ 5 \\ -1 \end{pmatrix}$$

$$= \begin{pmatrix} (3 \times 2)+ & (4 \times 5)+ & (0 \times -1) \\ (-2 \times 2)+ & (6 \times 5)+ & (-3 \times -1) \\ (7 \times 2)+ & (-4 \times 5)+ & (1 \times -1) \end{pmatrix} = \begin{pmatrix} 26 \\ 29 \\ -7 \end{pmatrix}$$

6   In algebra, the commutative law of multiplication states that $a \times b = b \times a$. For matrices, this law is only true in a few special cases, and in general $A \times B$ is not equal to $B \times A$.

7   A **unit matrix**, I, is one which all elements of the leading diagonal ($\searrow$) have a value of 1 and all other elements have a value of 0. Multiplication of a matrix by I is the equivalent of multiplying by 1 in arithmetic.

8   The **determinant** of a 2 by 2 matrix, $\begin{pmatrix} a & b \\ c & c \end{pmatrix}$ is defined as $(ad - bc)$. The elements of the determinant of a matrix are written between vertical lines. Thus, the determinant of $\begin{pmatrix} 3 & -4 \\ 1 & 6 \end{pmatrix}$ is written as $\begin{vmatrix} 3 & -4 \\ 1 & 6 \end{vmatrix}$ and is equal to $(3 \times 6) - (-4 \times 1)$, i.e. $18 - (-4)$ or 22. Hence the determinant of a matrix can be expressed as a single numerical value,

i.e. $\begin{vmatrix} 3 & -4 \\ 1 & 6 \end{vmatrix} = 22$

The inverse or reciprocal of a 2 by 2 matrix

9   The inverse of matrix A is A^{-1} such that $A \times A^{-1} = 1$, the unit matrix. For matrix $A = \begin{pmatrix} p & q \\ r & s \end{pmatrix}$ the inverse may be obtained by:

(i) interchanging the positions of p and s,
(ii) changing the signs of q and r, and
(iii) multiplying this new matrix by the reciprocal of the determinant of $\begin{pmatrix} p & q \\ r & s \end{pmatrix}$

Thus if $A = \begin{pmatrix} p & q \\ r & s \end{pmatrix}$ then $A^{-1} = \dfrac{1}{ps - qr} \begin{pmatrix} s & -q \\ -r & p \end{pmatrix}$

For example, the inverse of matrix $\begin{pmatrix} 1 & 2 \\ 3 & 4 \end{pmatrix}$ is

$$\frac{1}{4-6} \begin{pmatrix} 4 & -2 \\ -3 & 1 \end{pmatrix} = \begin{pmatrix} -2 & 1 \\ \dfrac{3}{2} & -\dfrac{1}{2} \end{pmatrix}$$

The determinant of a 3 by 3 matrix

10   (i) The **minor** of an element of a 3 by 3 matrix is the value of the 2 by 2 determinant obtained by covering up the row and column containing that element. Thus for the matrix

$$\begin{pmatrix} 1 & 2 & 3 \\ 4 & 5 & 6 \\ 7 & 8 & 9 \end{pmatrix}$$

the minor of element 4 is obtained by covering the row (4 5 6) and the column $\begin{pmatrix} 1 \\ 4 \\ 7 \end{pmatrix}$, leaving the 2 by 2 determinant

$$\begin{vmatrix} 2 & 3 \\ 8 & 9 \end{vmatrix},$$

i.e., the minor of element 4 is $(2 \times 9) - (3 \times 8)$, i.e. -6.

(ii) The sign of a minor depends on its position within the matrix, the sign pattern being

$$\begin{pmatrix} + & - & + \\ - & + & - \\ + & - & + \end{pmatrix}$$

Thus the signed-minor of element 4 in the matrix

$$\begin{pmatrix} 1 & 2 & 3 \\ 4 & 5 & 6 \\ 7 & 8 & 9 \end{pmatrix} \text{ is } -\begin{vmatrix} 2 & 3 \\ 8 & 9 \end{vmatrix} = -(-6) = 6.$$

The signed-minor of an element is called the **cofactor** of the element.

(iii) The value of a 3 by 3 determinant is the sum of the products of the elements and their cofactors **of any row or any column** of the corresponding 3 by 3 matrix.

For example,

$$\begin{vmatrix} a_1 & b_1 & c_1 \\ a_2 & b_2 & c_2 \\ a_3 & b_3 & c_3 \end{vmatrix} = a_1 \begin{vmatrix} b_2 & c_2 \\ b_3 & c_3 \end{vmatrix} - b_1 \begin{vmatrix} a_2 & c_2 \\ a_3 & c_3 \end{vmatrix} + c_1 \begin{vmatrix} a_2 & b_2 \\ a_3 & b_3 \end{vmatrix}$$

The inverse or reciprocal of a 3 by 3 matrix

11   The **adjoint** of a matrix A is obtained by:

(i) forming a matrix B of the cofactors of A, and

(ii) **transposing** matrix B to give B^T, where B^T is the matrix obtained by writing the rows of B as the columns of B^T. Then adj $A = B^T$. The inverse of matrix A, A^{-1} is given by $A^{-1} = \dfrac{\text{adj } A}{A}$, where adj A is the adjoint of matrix A and A is the determinant of matrix A.

Properties of determinants

12   There are certain properties of determinants which enable the value of a determinant to be found more simply. Some of these properties are given below.

(i) If all the elements in a row or column are interchanged with the corresponding elements in another row or column, the value of the determinant obtained is -1 times the value of the original determinant. Thus

$$\begin{vmatrix} a & b \\ c & d \end{vmatrix} = (-1) \times \begin{vmatrix} b & a \\ d & c \end{vmatrix}$$

(ii) If two rows or two columns of a determinant are equal its value is equal to zero. Thus:

$$\begin{vmatrix} a & a \\ c & c \end{vmatrix} = 0$$

(iii) If all the elements in any row or any column of a determinant have a common factor, the elements in that row or column can be divided by the common factor and the factor becomes a factor of the determinant. Thus

$$\begin{vmatrix} a & b \\ b \times x & b \times d \end{vmatrix} \equiv b \times \begin{vmatrix} a & b \\ c & d \end{vmatrix}$$

(iv) The value of a determinant remains unaltered if a multiple of the elements in any row or any column are added to the corresponding elements of any other row or column. Thus:

$$\begin{vmatrix} a & b \\ c & d \end{vmatrix} \equiv \begin{vmatrix} a & b + ka \\ c & d + kc \end{vmatrix}$$

The properties of determinants listed above may be used to reduce the size of elements within the determinant, and to introduce as many zero elements as is practical before evaluating the determinant. Simplification is mainly achieved by using property (iv)

Thus, for example, to simplify and evaluate $\begin{vmatrix} 2 & 7 & 26 \\ 1 & 2 & 6 \\ 4 & 11 & 40 \end{vmatrix}$

taking 3 times column 2 from column 3 gives:

$$\begin{vmatrix} 2 & 7 & (26-21) \\ 1 & 2 & (6-6) \\ 4 & 11 & (40-33) \end{vmatrix}$$

i.e. $\begin{vmatrix} 2 & 7 & 5 \\ 1 & 2 & 0 \\ 4 & 11 & 7 \end{vmatrix}$

Taking twice column 1 from column 2 gives $\begin{vmatrix} 2 & (7-4) & 5 \\ 1 & (2-2) & 0 \\ 4 & (11-8) & 7 \end{vmatrix}$

i.e. $\begin{vmatrix} 2 & 3 & 5 \\ 1 & 0 & 0 \\ 4 & 3 & 7 \end{vmatrix}$. Since two of the elements in row 2 are zero, the

value of this determinant is

$$-1 \times \begin{vmatrix} 3 & 5 \\ 3 & 7 \end{vmatrix} + 0 - 0 = -3 \times \begin{vmatrix} 1 & 5 \\ 1 & 7 \end{vmatrix}, \text{ (property (iii))}$$
$$= -3(7-5) = -\mathbf{6}$$

13 **The procedure for solving linear simultaneous equations in two unknowns using matrices is:**

(i) write the equations in the form

$a_1 x + b_1 y = c_1$
$a_2 x + b_2 y = c_2,$

(ii) write the matrix equation corresponding to these equations,

i.e., $\begin{pmatrix} a_1 & b_1 \\ a_2 & b_2 \end{pmatrix} \times \begin{pmatrix} x \\ y \end{pmatrix} = \begin{pmatrix} c_1 \\ c_2 \end{pmatrix},$

(iii) determine the inverse matrix of $\begin{pmatrix} a_1 & b_1 \\ a_2 & b_2 \end{pmatrix},$

i.e., $\dfrac{1}{a_1 b_2 - b_1 a_2} \begin{pmatrix} b_2 & -b_1 \\ -a_2 & a_1 \end{pmatrix},$ (from para 9),

(iv) multiply each side of (ii) by the inverse matrix, and

(v) solve for x and y by equating corresponding elements.

For example, to solve the simultaneous equations

$$3x + 5y - 7 = 0 \qquad (1)$$
$$4x - 3y - 19 = 0 \qquad (2)$$

using matrices, following the above procedure:

(i) Writing the equations in the $a_1 x + b_1 y = c_1$ form gives:

$$3x + 5y = 7$$
$$4x - 3y = 19$$

(ii) The matrix equation is

$$\begin{pmatrix} 3 & 5 \\ 4 & -3 \end{pmatrix} \times \begin{pmatrix} x \\ y \end{pmatrix} = \begin{pmatrix} 7 \\ 19 \end{pmatrix}$$

(iii) The inverse of matrix $\begin{pmatrix} 3 & 5 \\ 4 & -3 \end{pmatrix}$ is

$$\frac{1}{3 \times (-3) - 5 \times 4} \begin{pmatrix} -3 & -5 \\ -4 & 3 \end{pmatrix} \text{ i.e.} \begin{pmatrix} \dfrac{3}{29} & \dfrac{5}{29} \\ \dfrac{4}{29} & \dfrac{-3}{29} \end{pmatrix}$$

(iv) Multiplying each side of (ii) by (iii) and remembering that $A \times A^{-1} = I$, the unit matrix, gives:

$$\begin{pmatrix} 1 & 0 \\ 0 & 1 \end{pmatrix} \begin{pmatrix} x \\ y \end{pmatrix} = \begin{pmatrix} \dfrac{3}{29} & \dfrac{5}{29} \\ \dfrac{4}{29} & \dfrac{-3}{29} \end{pmatrix} \times \begin{pmatrix} 7 \\ 19 \end{pmatrix}$$

Thus $\begin{pmatrix} x \\ y \end{pmatrix} = \begin{pmatrix} \dfrac{21}{29} + \dfrac{95}{29} \\ \dfrac{28}{29} - \dfrac{57}{29} \end{pmatrix}$

i.e. $\begin{pmatrix} x \\ y \end{pmatrix} = \begin{pmatrix} 4 \\ -1 \end{pmatrix}$

(v) By comparing corresponding elements,

$x = 4$ and $y = -1$.

$$\left[\begin{array}{l} \text{Checking: Equation (1), } 3 \times 4 + 5 \times (-1) - 7 = 0 = \text{RHS} \\ \qquad \text{Equation (2), } 4 \times 4 - 3 \times (-1) - 19 = 0 = \text{RHS} \end{array} \right]$$

75

14 **The procedure for solving linear simultaneous equations in three unknowns by matrices is:**

(i) Write the equations in the form

$a_1x + b_1y + c_1z = d_1$
$a_2x + b_2y + c_2z = d_2$
$a_3x + b_3y + c_3z = d_3$

(ii) Write the matrix equation corresponding to these equations, i.e.

$$\begin{pmatrix} a_1 & b_1 & c_1 \\ a_2 & b_2 & c_2 \\ a_3 & b_3 & c_3 \end{pmatrix} \times \begin{pmatrix} x \\ y \\ z \end{pmatrix} = \begin{pmatrix} d_1 \\ d_2 \\ d_3 \end{pmatrix}$$

(iii) Determine the inverse matrix of

$$\begin{pmatrix} a_1 & b_1 & c_1 \\ a_2 & b_2 & c_2 \\ a_3 & b_3 & c_3 \end{pmatrix}$$

(see para 11),

(iv) Multiply each side of (ii) by the inverse matrix, and
(v) Solve for x, y and z by equating the corresponding elements. For example, to solve the simultaneous equations

$x + y + z - 4 = 0$ (1)
$2x - 3y + 4z - 33 = 0$ (2)
$3x - 2y - 2z - 2 = 0$ (3)

using matrices following the above procedure:

(i) Writing the equations in the $a_1x + b_1y + c_1z = d_1$ form gives:

$x + y + z = 4$
$2x - 3y + 4z = 33$
$3x - 2y - 2z = 2$

(ii) The matrix equation is

$$\begin{pmatrix} 1 & 1 & 1 \\ 2 & -3 & 4 \\ 3 & -2 & -2 \end{pmatrix} \times \begin{pmatrix} x \\ y \\ z \end{pmatrix} = \begin{pmatrix} 4 \\ 33 \\ 2 \end{pmatrix}$$

(iii) The inverse matrix of $A = \begin{pmatrix} 1 & 1 & 1 \\ 2 & -3 & 4 \\ 3 & -2 & -2 \end{pmatrix}$

is given by $A^{-1} = \dfrac{\text{adj } A}{A}$

The adjoint of A is the transpose of the matrix of the cofactors of the elements, (see para 11). The matrix of cofactors is

$$\begin{pmatrix} 14 & 16 & 5 \\ 0 & -5 & 5 \\ 7 & -2 & -5 \end{pmatrix}$$

and the transpose of this matrix gives

$$\text{adj } A = \begin{pmatrix} 14 & 0 & 7 \\ 16 & -5 & -2 \\ 5 & 5 & -5 \end{pmatrix}$$

The determinant of A, i.e., the sum of the products of elements and their cofactors, using a first row expansion is

$$1\begin{vmatrix} -3 & 4 \\ -2 & -2 \end{vmatrix} - 1\begin{vmatrix} 2 & 4 \\ 3 & -2 \end{vmatrix} + 1\begin{vmatrix} 2 & -3 \\ 3 & -2 \end{vmatrix}$$

i.e. $(1 \times 14) - (1 \times -16) + (1 \times 5)$, that is, 35. Hence, the inverse of A,

$$A^{-1} = \frac{1}{35}\begin{pmatrix} 14 & 0 & 7 \\ 16 & -5 & -2 \\ 5 & 5 & -5 \end{pmatrix}$$

(iv) Multiplying each side of (ii) by (iii), remembering that $A \times A^{-1} = I$, the unit matrix, gives

$$\begin{pmatrix} 1 & 0 & 0 \\ 0 & 1 & 0 \\ 0 & 0 & 1 \end{pmatrix} \times \begin{pmatrix} x \\ y \\ z \end{pmatrix} = \frac{1}{35}\begin{pmatrix} 14 & 0 & 7 \\ 16 & -5 & -2 \\ 5 & 5 & -5 \end{pmatrix} \times \begin{pmatrix} 4 \\ 33 \\ 2 \end{pmatrix}$$

$$\begin{pmatrix} x \\ y \\ z \end{pmatrix} = \frac{1}{35}\begin{pmatrix} (14 \times 4) + (0 \times 33) + (7 \times 2) \\ (16 \times 4) + (-5 \times 33) + ((-2) \times 2) \\ (5 \times 4) + (5 \times 33) + ((-5) \times 2) \end{pmatrix}$$

$$= \frac{1}{35}\begin{pmatrix} 70 \\ -105 \\ 175 \end{pmatrix} = \begin{pmatrix} 2 \\ -3 \\ 5 \end{pmatrix}$$

(v) By comparing corresponding elements, $x = 2$, $y = -3$, $z = 5$ which can be checked in the original equations.

15 **When solving linear simultaneous equations in two unknowns using determinants:**

(i) write the equations in the form

$$a_1 x + b_1 y + c_1 = 0$$
$$a_2 x + b_2 y + c_2 = 0$$

and then:

(ii) the solution is given by

$$\frac{x}{D_x} = \frac{-y}{D_y} = \frac{1}{D}$$

where

$$D_x = \begin{vmatrix} b_1 & c_1 \\ b_2 & c_2 \end{vmatrix}$$

i.e. the determinant of the coefficients left when the x-column is covered up,

$$D_y = \begin{vmatrix} a_1 & c_1 \\ a_2 & c_2 \end{vmatrix}$$

i.e. the determinant of the coefficients left when the y-column is covered up, and

$$D = \begin{vmatrix} a_1 & b_1 \\ a_2 & b_2 \end{vmatrix}$$

i.e. the determinant of the coefficients left when the constants-column is covered up.

For example, to solve the simultaneous equations:

$$3x + 4y = 0$$
$$2x + 5y + 7 = 0$$

the solution is given by:

$$\frac{x}{\begin{vmatrix} 4 & 0 \\ 5 & 7 \end{vmatrix}} = \frac{-y}{\begin{vmatrix} 3 & 0 \\ 2 & 7 \end{vmatrix}} = \frac{1}{\begin{vmatrix} 3 & 4 \\ 2 & 5 \end{vmatrix}} .$$

i.e. $\dfrac{x}{28} = \dfrac{-y}{21} = \dfrac{1}{7}$

from which,

$$x = \frac{28}{7} = \mathbf{4} \text{ and } y = \frac{-21}{7} = \mathbf{-3}$$

16   When solving simultaneous equations in three unknowns using determinants:

(i) Write the equations in the form

$$a_1x + b_1y + c_1z + d_1 = 0$$
$$a_2x + b_2y + c_2z + d_2 = 0$$
$$a_3x + b_3y + c_3z + d_3 = 0$$

and then

(ii) the solution is given by

$$\frac{x}{D_x} = \frac{-y}{D_y} = \frac{z}{D_z} = \frac{-1}{D}$$

where

$$D_x \text{ is } \begin{vmatrix} b_1 & c_1 & d_1 \\ b_2 & c_2 & d_2 \\ b_3 & c_3 & d_3 \end{vmatrix}$$

i.e. the determinant of the coefficients obtained by covering up the x-column,

$$D_y \text{ is } \begin{vmatrix} a_1 & c_1 & d_1 \\ a_2 & c_2 & d_2 \\ a_3 & c_3 & d_3 \end{vmatrix}$$

i.e. the determinant of the coefficients obtained by covering up the y-column,

$$D_z \text{ is } \begin{vmatrix} a_1 & b_1 & d_1 \\ a_2 & b_2 & d_2 \\ a_3 & b_3 & d_3 \end{vmatrix}$$

i.e. the determinant of the coefficients obtained by covering up the z-column, and

$$D \text{ is } \begin{vmatrix} a_1 & b_1 & c_1 \\ a_2 & b_2 & c_2 \\ a_3 & b_3 & c_3 \end{vmatrix}$$

i.e. the determinant of the coefficients obtained by covering up the constants-column.

For example, to solve

$$2x + 3y - 4z - 26 = 0$$
$$x - 5y - 3z + 87 = 0$$
$$-7x + 2y + 6z - 12 = 0$$

79

the solution is given by:

$$\frac{x}{\begin{vmatrix} 3 & -4 & -26 \\ -5 & -3 & 87 \\ 2 & 6 & -12 \end{vmatrix}} = \frac{-y}{\begin{vmatrix} 2 & -4 & -26 \\ 1 & -3 & 87 \\ -7 & 6 & -12 \end{vmatrix}}$$

$$= \frac{z}{\begin{vmatrix} 2 & 3 & -26 \\ 1 & -5 & 87 \\ -7 & 2 & -12 \end{vmatrix}} = \frac{-1}{\begin{vmatrix} 2 & 3 & -4 \\ 1 & -5 & -3 \\ -7 & 2 & 6 \end{vmatrix}}$$

i.e. $\dfrac{x}{-1290} = \dfrac{-y}{1806} = \dfrac{z}{-1161} = \dfrac{-1}{129}$

from which,

$$x = \frac{-1290}{-129} = \mathbf{10}, \; y = \frac{-1860}{-129} = \mathbf{14} \text{ and } z = \frac{-1161}{-129} = \mathbf{9}$$

6 Complex numbers

1   If the quadratic equation $x^2 + 2x + 5 = 0$ is solved using the quadratic formula then

$$x = \frac{-2 \pm \sqrt{[(2)^2 - (4)(1)(5)]}}{2(1)} = \frac{-2 \pm \sqrt{-16}}{2} = \frac{-2 \pm \sqrt{[(16)(-1)]}}{2}$$

$$= \frac{-2 \pm \sqrt{16}\sqrt{-1}}{2} = \frac{-2 \pm 4\sqrt{-1}}{2}$$

$$= -1 \pm 2\sqrt{-1}$$

It is not possible to evaluate $\sqrt{-1}$ in real terms. However, if an operator j is defined as $j = \sqrt{-1}$ then the solution may be expressed as $x = -1 \pm j2$.

2    $-1 + j2$ and $-1 - j2$ are known as **complex numbers**. Both solutions are of the form $a + jb$, 'a' being termed the **real part** and jb the **imaginary part**. A complex number of the form $a + jb$ is called a **cartesian complex number**.

3   In pure mathematics the symbol i is used to indicate $\sqrt{-1}$ (i being the first letter of the word imaginary). However i is the symbol of electric current, and to avoid possible confusion the next letter in the alphabet, j, is used to represent $\sqrt{-1}$.

4   A complex number may be represented pictorially on rectangular or cartesian axes. The horizontal (or x) axis is used to represent the real axis and the vertical (or y) axis is used to represent the imaginary axis. Such a diagram is called an **Argand diagram**. In *Figure 6.1*, the point A represents the complex number $(3 + j2)$ and is obtained by plotting the coordinates $(3, j2)$ as in graphical work. *Figure 6.1* also shows the Argand points B, C and D representing the complex numbers $(-2 + j4)$, $(-3 - j5)$ and $(1 - j3)$ respectively.

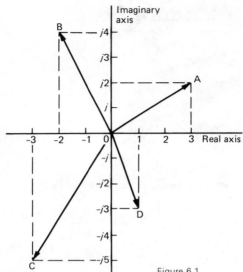

Figure 6.1

Addition and subtraction of complex numbers

5   Two complex numbers are added/subtracted by adding/subtracting separately the two real parts and the two imaginary parts. Thus, for example,

$$(2 + j3) + (3 - j4) = 2 + j3 + 3 - j4 = 5 - j1$$

and

$$(2 + j3) - (3 - j4) = 2 + j3 - 3 + j4 = -1 + j7.$$

Multiplication and division of complex numbers

6   **Multiplication of complex numbers** is achieved by assuming all quantities involved are real and then using $j^2 = -1$ to simplify. Thus

$$(3 + j2)(4 - j5) = 12 - j15 + j8 - j^2 10 = (12 - -10) + j(-15 + 8)$$
$$= 22 - j7$$

7   The **complex conjugate** of a complex number is obtained by changing the sign of the imaginary part. Hence the complex conjugate of $a + jb$ is $a - jb$. The product of a complex number and its complex conjugate is always a real number. For example,

$$(3 + j4)(3 - j4) = 9 - j12 + j12 - j^2 16 = 9 + 16 = 25.$$

[$(a + jb)(a - jb)$ may be evaluated 'on sight' as $a^2 + b^2$].

8   **Division of complex numbers** is achieved by multiplying both numerator and denominator by the complex conjugate of the denominator. For example,

$$\frac{2 - j5}{3 + j4} = \frac{2 - j5}{3 + j4} \times \frac{(3 - j4)}{(3 - j4)} = \frac{6 - j8 - j15 + j^2 20}{3^2 + 4^2} = \frac{-14 - j23}{25}$$

$$= \frac{-14}{25} - j\frac{23}{25} \text{ or } -0.56 - j0.92$$

Complex equations

9   If two complex numbers are equal, then their real parts are equal and their imaginary parts are equal. Hence if $a + jb = c + jd$, then $a = c$ and $b = d$.

The polar form of a complex number

(i) Let a complex number Z be $x + jy$, as shown in the Argand diagram of *Figure 6.2*. Let distance OZ be r and the angle OZ makes with the positive real axis be θ. From trigonometry, $x = r \cos \theta$ and $y = r \sin \theta$. Hence $Z = x + jy = r \cos \theta + jr \sin \theta$ $= r(\cos \theta + j \sin \theta)$.
$Z = r(\cos \theta + j \sin \theta)$ is usually abbreviated to $Z = r \angle \theta$ which is known as the polar form of a complex number.

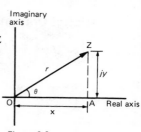

Figure 6.2

(ii) r is called the modulus (or magnitude) of Z and is written as mod Z or $|Z|$, r is determined using Pythagoras' theorem on triangle OAZ in fig. 2, i.e., $r = \sqrt{(x^2 + y^2)}$.
(iii) θ is called the argument (or amplitude) of Z and is written as arg Z. By trigonometry on triangle OAZ,

$$\text{arg } Z = \theta = \arctan \frac{y}{x}.$$

(iv) Whenever changing from cartesian form to polar form, or vice-versa, a sketch is invaluable for determining the quadrant in which the complex number occurs. Thus, for example, to determine $(3+j4)$ in $r\angle\theta$ form,

$$r = \sqrt{(3^2+4^2)} = 5$$

and, from *Figure 6.3(a)*, $\theta = \arctan\dfrac{4}{3} = 53°8'$.

Hence $(3+j4) = 5\angle 53°8'$.

Also $12\angle 165° = 12\cos 165° + j12\sin 165°$
$$= -11.59 + j3.11,\text{ as shown in } \textit{Figure 6.3(b)}.$$

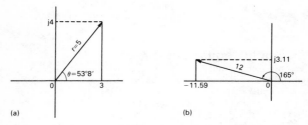

(a)   (b)

Figure 6.3

Multiplication and division in polar form

11   If $Z_1 = r_1\angle\theta_1$ and $Z_2 = r_2\angle\theta_2$ then:
        (i) $Z_1 Z_2 = r_1 r_2\angle(\theta_1+\theta_2)$, and

        (ii) $\dfrac{Z_1}{Z_2} = \dfrac{r_1}{r_2}\angle(\theta_1-\theta_2)$.

Thus, for example,

$$8\angle 25° \times 4\angle 60° = (8\times 4)\angle(25°+60°) = \mathbf{32\angle 85°}$$

$$\text{and } \frac{16\angle 75°}{2\angle 15°} = \frac{16}{2}\angle(75°-15°) = \mathbf{8\angle 60°}$$

De Moivre's theorem

12   **De Moivre's theorem** states:

$$[r\angle\theta]^n = r^n\angle n\theta,$$

which is true for all positive, negative or fractional values of n, and is thus useful for determining powers and roots of complex numbers. For example,

$$[3 \angle 20°]^4 = 3^4 \angle (4 \times 20°) = 81 \angle 80°$$

The **square root of a complex number** is determined by letting $n = \dfrac{1}{2}$ in De Moivre's theorem, i.e.,

$$\sqrt{[r \angle \theta]} = [r \angle \theta]^{1/2} = r^{1/2} \angle \frac{1}{2}\theta = \sqrt{r} \angle \frac{\theta}{2}.$$

There are two square roots of a real number, equal in size but opposite in sign. Thus, for example, to find the two square roots of $(5 + j12)$:

$$(5 + j12) = \sqrt{(5^2 + 12^2)} \angle \arctan \frac{12}{5} = 13 \angle 67° 23'$$

When determining square roots two solutions result. To obtain the second solution one way is to express $13 \angle 67°23'$ also as $13 \angle (67°23' + 360°)$, i.e., $13 \angle 427°23'$. When the angle is divided by 2 an angle less than $360°$ is obtained. Hence

$$\begin{aligned}
\sqrt{(5 + j12)} &= \sqrt{[13 \angle 67°23']} \text{ and } \sqrt{[13 \angle 427°23']} \\
&= [13 \angle 67°23']^{1/2} \text{ and } [13 \angle 427°23']^{1/2} \\
&= 13^{1/2} \angle \left(\frac{1}{2} \times 67°23'\right) \text{ and } 13^{1/2} \angle \left(\frac{1}{2} \times 427°23'\right) \\
&= \sqrt{13} \angle 33°42' \text{ and } \sqrt{13} \angle 213°42' \\
&= 3.61 \angle 33°42' \text{ and } 3.61 \angle 213°42'
\end{aligned}$$

Thus, in polar form, the two roots are 3.61 $\angle$ 33°42' and 3.61 $\angle$ − 146° 18'.

$$3.61 \angle 33°42' = 3.61(\cos 33°42' + j \sin 33°42') = 3.0 + j2.0$$
$$\begin{aligned}
3.61 \angle -146°18' &= 3.61[\cos(-146°18') + j\sin(-146°18')] \\
&= -3.0 - j2.0
\end{aligned}$$

Thus, in cartesian form the two roots are $\pm(3.0 + j2.0)$.

From the Argand diagram shown in *Figure 6.4* the two roots are seen to be $180°$ apart, which is always true when finding square roots of complex numbers.

13   When finding the **n^{th} root of a complex number**, there are n solutions. For example, there are three solutions to a cube root, five solutions to a fifth root, and so on. In the solutions to the roots of a complex number, the modulus, r, is always the same, but the arguments, θ, are different. Arguments are symmetrically spaced on an Argand diagram and are $\left(\dfrac{360}{n}\right)^°$ apart, where n is the

85

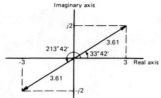

Figure 6.4

number of the root required. Thus if one of the solutions to the cube root of a complex number is, say, $5 \angle 20°$, the other two roots are symmetrically spaced $\left(\dfrac{360}{3}\right)°$, i.e., $120°$ from this root and the three roots are $5 \angle 20°$, $5 \angle 140°$ and $5 \angle 260°$. For example, to determine the roots of $(-14+j3)^{-2/5}$:

$$-14+j3 = 14.32 \angle 167° 54'$$

and

$$(-14+j3)^{-2/5} = 14.32^{-2/5} \angle \left(-\frac{2}{5} \times 167° 54'\right) = 0.3448 \angle -67° 10'$$

There are five roots each symmetrically placed around the Argand diagram, $\left(\dfrac{360}{5}\right)°$, i.e., $72°$ apart. Thus the roots are:

$0.3448 \angle -67° 10'$, $0.3448 \angle 4° 50'$, $0.3448 \angle 76° 50'$,
$0.3448 \angle 148° 50'$ and $0.3448 \angle 220° 50'$

Application of complex numbers

14   There are several applications of complex numbers in science and engineering, in particular in electrical alternating current theory and in mechanical vector analysis.

The effect of multiplying a phasor by j is to rotate it in a positive direction (i.e. anticlockwise) on an Argand diagram through $90°$ without altering its length. Similarly, multiplying a phasor by $-j$ rotates the phasor through $-90°$. These facts are used in a.c. theory since certain quantities in the phasor diagrams lie at $90°$ to each other. For example, in the R-L series circuit shown in *Figure 6.5(a)*, V_L leads I by $90°$ (i.e., I lags V_L by $90°$) and may be written as jV_L, the vertical axis being regarded as the imaginary axis of an Argand diagram. Thus $V_R + jV_L = V$ and since

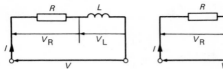
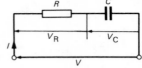

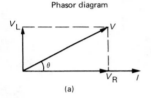
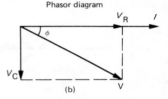

Figure 6.5

$V_R = IR$, $V = IX_L$ (where X_L is the inductive reactance, $2\pi fL$ ohms) and $V = IZ$ (where Z is the impedance) then $R + jV_L = Z$. Similarly, for the R-C circuit shown in *Figure 6.5(b)*, V_C lags I by 90° (i.e. I leads V_C by 90°) and $V_R - jV_C = V$, from which $R - jX_C = Z$ $\left(\text{where } X_C \text{ is the capacitive reactance, } \dfrac{1}{2\pi fC} \text{ ohms}\right)$.

Thus $Z = (4 + j7)\Omega$ represents a impedance consisting of a 4Ω resistance in series with an inducatance of inductive reactance 7Ω.

Similarly an impedance of $(5 - j3)\Omega$ represents a 5Ω resistance in series with a capacitor of capacitive reactance 3Ω. Complex numbers are particularly useful with parallel a.c. circuits. For example, for the circuit shown in *Figure 6.6* impedance Z for the three-branch parallel circuit is given by:

$$\frac{1}{Z} = \frac{1}{Z_1} + \frac{1}{Z_2} + \frac{1}{Z_3}, \text{ where } Z_1 = (4 + j3), Z_2 = 10, \text{ and}$$

$$Z_3 = 12 - j5.$$

Admittance, $Y_1 = \dfrac{1}{Z_1} = \dfrac{1}{4 + j3} = \dfrac{1}{4 + j3} \times \dfrac{4 - j3}{4 - j3} = \dfrac{4 - j3}{4^2 + 3^2}$
$$= 0.160 - j0.120 \text{ siemens.}$$

Admittance, $Y_2 = \dfrac{1}{Z_2} = \dfrac{1}{10} = 0.10 \text{ siemens.}$

87

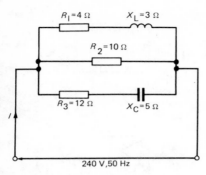

Figure 6.6

Admittance, $Y_3 = \dfrac{1}{Z_3} = \dfrac{1}{12 - j5} = \dfrac{1}{12 - j5} \times \dfrac{12 + j5}{12 + j5} = \dfrac{12 + j5}{12^2 + 5^2}$
$\qquad\qquad = 0.0710 + j0.0296$ siemens

Total admittance
$Y = Y_1 + Y_2 + Y_3 = (0.160 - j0.120) + (0.10) + (0.0710 + j0.0296)$
$\qquad\qquad\qquad\qquad = 0.331 - j0.0904$
$\qquad\qquad\qquad\qquad = 0.343 \angle -15° 17'$ siemens

Current $I = \dfrac{V}{Z} = VY = (240 \angle 0°)(0.343 \angle -15° 17')$
$= 82.32 \angle -15° 17'$ amperes.

Exponential form of a complex number

15 Certain mathematical functions may be expressed as power
series, three examples being:

$$(i) \quad e^x = 1 + x + \frac{x^2}{2!} + \frac{x^3}{3!} + \frac{x^4}{4!} + \frac{x^5}{5!} + \qquad (1)$$

$$(ii) \quad \sin x = x - \frac{x^3}{3!} + \frac{x^5}{5!} - \frac{x^7}{7!} + \qquad (2)$$

$$(iii) \quad \cos x = 1 - \frac{x^2}{2!} + \frac{x^4}{4!} - \frac{x^6}{6!} + \qquad (3)$$

16 Replacing x in equation (1) by the imaginary number $j\theta$, gives:

$$e^{j\theta} = 1 + j\theta + \frac{(j\theta)^2}{2!} + \frac{(j\theta)^3}{3!} + \frac{(j\theta)^4}{4!} + \frac{(j\theta)^5}{5!} + \ldots$$

$$= 1 + j\theta + \frac{j^2\theta^2}{2!} + \frac{j^3\theta^3}{3!} + \frac{j^4\theta^4}{4!} + \frac{j^5\theta^5}{5!} + \ldots$$

By definition, $j = \sqrt{(-1)}$, hence $j^2 = -1$, $j^3 = -j$, $j^4 = 1$, $j^5 = j$, and so on.

Thus, $e^{j\theta} = 1 + j\theta - \frac{\theta^2}{2!} - j\frac{\theta^3}{3!} + \frac{\theta^4}{4!} + j\frac{\theta^5}{5!} - \ldots$

Grouping real and imaginary terms gives:

$$e^{j\theta} = \left(1 - \frac{\theta^2}{2!} + \frac{\theta^4}{4!} - \ldots\right) + j\left(\theta - \frac{\theta^3}{3!} + \frac{\theta^5}{5!} - \ldots\right)$$

However, from equations (2) and (3):

$$\left(1 - \frac{\theta^2}{2!} + \frac{\theta^4}{4!} - \ldots\right) = \cos\theta \text{ and } \left(\theta - \frac{\theta^3}{3!} + \frac{\theta^5}{5!} - \ldots\right) = \sin\theta$$

Thus $e^{j\theta} = \cos\theta + j\sin\theta$ (4)

Writing $-\theta$ for θ in equation (4), gives:

$$e^{j(-\theta)} = \cos(-\theta) + j\sin(-\theta)$$

However, $\cos(-\theta) = \cos\theta$ and $\sin(-\theta) = -\sin\theta$

Thus $e^{-j\theta} = \cos\theta - j\sin\theta$ (5)

17   (i) The polar form of a complex number z is:
r$(\cos\theta + j\sin\theta)$. But, from equation (4),
$\cos\theta + j\sin\theta = e^{j\theta}$. Therefore, $z = re^{j\theta}$. When a complex number is written in this way, it is said to be expressed in **exponential form**.

(ii) The exponential form of a complex number is required when finding **logarithms of complex numbers**. For example

$$\ln(3 + j4) = \ln[5\angle 53°8'] = \ln[5\angle 0.927]$$
$$\text{since } 53°8' = 0.927 \text{ rad}$$
$$= \ln[5e^{j0.927}]$$
$$= \ln 5 + \ln e^{j0.927} \text{ from the laws of logarithms}$$
$$= \ln 5 + j0.927, \text{ from the definition of a logarithm}$$

i.e. $\mathbf{\ln(3 + j4) = 1.609 + j0.927}$ correct to 3 decimal places.

Expressing cos $n\theta$ and sin $n\theta$ in terms of powers of cos θ and sin θ

18   The complex number $z = r(\cos \theta + j \sin \theta)$, when raised to power n, becomes $z^n = r^n (\cos n\theta + j \sin n\theta)$ by de Moirvre's theorem.

Also $z^n = [r(\cos \theta + j \sin \theta)]^n$
$= r^n(\cos \theta + j \sin \theta)^n$

Thus, $\cos n\theta + j \sin n\theta = (\cos \theta + j \sin \theta)^n$ $\qquad\qquad$ (6)

The right-hand side of this equation can be expanded by applying the binomial theorem, i.e.

$$(\cos \theta + j \sin \theta)^n = \cos^n\theta + n \cos^{n-1}\theta j \sin \theta + \frac{n(n-1)}{2!}\cos^{n-2}\theta j^2 \sin^2\theta +$$

But $j^2 = -1$, $j^3 = -j$, $j^4 = 1$, and so on.

Hence, $(\cos \theta + j \sin \theta)^n = \cos^n\theta + jn \cos^{n-1} \theta \sin \theta -$

$$\frac{n(n-1)}{2!} \cos^{n-2} \theta \sin^2 \theta - j\frac{n(n-1)(n-2)}{3!} \cos^{n-3}\theta \sin^3\theta +$$

Grouping real and imaginary terms gives:

$$(\cos \theta + j \sin \theta)^n = \cos^n\theta - \frac{n(n-1)}{2!}\cos^{n-2}\theta \sin^2\theta +$$

$$\frac{n(n-1)(n-2)(n-3)}{4!} \cos^{n-4}\theta \sin^4\theta - \ldots$$

$$+ j\left(n \cos^{n-1}\theta \sin \theta - \frac{n(n-1)(n-2)}{3!} \cos^{n-3}\theta \sin^3\theta + \ldots \right)$$

and from equation (6):

$$(\cos n\theta + j \sin n\theta) = \cos^n\theta - \frac{n(n-1)}{2!} \cos^{n-2}\theta \sin^2\theta +$$

$$\frac{n(n-1)(n-2)(n-3)}{4!} \cos^{n-4} \sin^4\theta - \ldots$$

$$+ j\left(n \cos^{n-1}\theta \sin \theta - \frac{n(n-1)(n-2)}{3!} \cos^{n-3} \theta \sin^3\theta + \ldots \right)$$

Equating the real parts:

$$\cos n\theta = \cos^n\theta - \frac{n(n-1)}{2!} \cos^{n-2}\theta \sin^2\theta + \ldots \qquad (7)$$

Equating the imaginary parts:

$$\sin n\theta = n \cos^{n-1}\theta \sin \theta - \frac{n(n-1)(n-2)}{3!} \cos^{n-3}\theta \sin^3\theta + \dots \quad (8)$$

Equations (7) and (8) are used to express $\cos n\theta$ and $\sin n\theta$ in terms of powers of $\cos \theta$ and $\sin \theta$.

Thus, for example, from equation (7):

$$\cos 4\theta = \cos^4\theta - \frac{4(3)}{2!} \cos^2\theta \sin^2\theta + \frac{(4)(3)(2)(1)}{4!} \cos^0\theta \sin^4\theta$$

$$= \cos^4\theta - 6 \cos^2\theta \sin^2\theta + \sin^4\theta$$

and from equation (8):

$$\sin 5\theta = 5 \cos^4\theta \sin \theta - \frac{(5)(4)(3)}{2!} \cos^2\theta \sin^3\theta +$$

$$\frac{(5)(4)(3)(2)(1)}{5!} \cos^0\theta \sin^5\theta$$

$$= 5 \cos^4\theta \sin \theta - 10 \cos^2\theta \sin^3\theta + \sin^5\theta$$

Expressing $\cos^n\theta$ and $\sin^n\theta$ in terms of sines and cosines of multiples of θ

19   From para 16, equations (4) and (5):

$$z = e^{j\theta} = \cos \theta + j \sin \theta \quad (9)$$

$$\frac{1}{z} = \frac{1}{e^{j\theta}} = e^{-j\theta} = \cos \theta - j \sin \theta \quad (10)$$

Adding equations (9) and (10) gives:

$$z + \frac{1}{z} = 2 \cos \theta$$

and it follows that

$$\left(z + \frac{1}{z}\right)^n = 2^n \cos^n\theta \quad (11)$$

Subtracting equation (10) from equation (9) gives:

$$z - \frac{1}{z} = j 2 \sin \theta$$

It follows that

$$\left(z - \frac{1}{z}\right)^n = j^n 2^n \sin^n\theta \quad (12)$$

If $z = \cos\theta + j\sin\theta$, then from de Moivre's theorem,

$$z^n = \cos n\theta + j\sin n\theta$$

Also, if $\dfrac{1}{z} = \cos\theta - j\sin\theta$, then from de Moivre's theorem,

$$\frac{1}{z^n} = \cos n\theta - j\sin n\theta$$

Adding gives

$$\left(z^n + \frac{1}{z^n}\right) = 2\cos n\theta \tag{13}$$

Subtracting gives

$$\left(z^n - \frac{1}{z^n}\right) = 2j\sin n\theta \tag{14}$$

Equations (11) to (14) are used for expressing powers of $\cos\theta$ and $\sin\theta$ in terms of cosines and sines of multiples of θ.

For example, to express $\sin^2 C$ in terms of cosines of multiples of C:

From equation (12), $j^n 2^n \sin^n\theta = \left(z - \dfrac{1}{z}\right)^n$

When $n = 2$, $j^2 2^2 \sin^2 C = \left(z - \dfrac{1}{z}\right)^2 = z^2 - 2 + \dfrac{1}{z^2} = \left(z^2 + \dfrac{1}{z^2}\right) - 2$

From equation (13), when $n = 2$, $\left(z^2 + \dfrac{1}{z^2}\right) = 2\cos 2C$.

Hence $j^2 2^2 \sin^2 C = 2\cos 2C - 2$, and since $j^2 = 1$,
$-4\sin^2 C = 2\cos 2C - 2$

$\sin^2 C = \dfrac{1}{4}(2 - 2\cos 2C)$, i.e., $\mathbf{\sin^2 C = \dfrac{1}{2}(1 - \cos 2C)}$

Similarly to express $\cos^5\theta$ in terms of cosines of multiples of θ:

From equation (11), $2^n \cos^n\theta = \left(z^1 + \dfrac{1}{z}\right)^n$

Expanding $\left(z + \dfrac{1}{z}\right)^5$ by the binomial theorem gives:

$$\left(z + \frac{1}{z}\right)^5 = z^5 + 5z^4 \cdot \frac{1}{z} + 10z^3 \cdot \frac{1}{z^2} + 10z^2 \cdot \frac{1}{z^3} + 5z \cdot \frac{1}{z^4} + \frac{1}{z^5}$$

$$= z^5 + 5z^3 + 10z + \frac{10}{z} + \frac{5}{z^3} + \frac{1}{z^5}$$

$$= \left(z^5 + \frac{1}{z^5}\right) + 5\left(z^3 + \frac{1}{z^3}\right) + 10\left(z + \frac{1}{z}\right)$$

But from equation (13), $z^n + \dfrac{1}{z^n} = 2\cos n\theta$

Hence $\left(z + \dfrac{1}{z}\right)^5 = 2\cos 5\theta + 5(2\cos 3\theta) + 10(2\cos\theta)$

Thus $2^5\cos^5\theta = 2\cos 5\theta + 10\cos 3\theta + 20\cos\theta$

i.e. $\cos^5\theta = \dfrac{1}{2^4}(\cos 5\theta + 5\cos 3\theta + 10\cos\theta)$

7 Geometry

1   Geometry is a part of mathematics in which the properties of points, lines, surfaces and solids are investigated.

Angles

2   An angle is the amount of rotation between two straight lines. Angles may be measured in either degrees or radians (see para 15).

1 revolution = 360 degrees, thus 1 degree $= \frac{1}{360}$th of one revolution.

Also 1 minute $= \frac{1}{60}$th of a degree and 1 second $= \frac{1}{60}$th of a minute.
1 minute is written as 1′ and 1 second is written as 1″.
Thus $1° = 60'$ and $1' = 60''$.

3   (i) Any angle between 0° and 90° is called an **acute angle**.
          (ii) An angle equal to 90° is called a **right angle**.
          (iii) Any angle between 90° and 180° is called an **obtuse angle**.
          (iv) Any angle greater than 180° and less than 360° is called a **reflex angle**.

4   (i) An angle of 180° lies on a straight line.
          (ii) If two angles add up to 90° they are called **complementary angles**.
          (iii) If two angles add up to 180° they are called **supplementary angles**.
          (iv) **Parallel lines** are straight lines which are in the same plane and never meet. (Such lines are denoted by arrows, as in *Figure 7.1*).
          (v) A straight line which crosses two parallel lines is called a **transversal** (see MN in *Figure 7.1*).

5   With reference to *Figure 7.1*:
          (i) $a=c$, $b=d$, $e=g$ and $f=h$. Such pairs of angles are called **vertically opposite angles**.
          (ii) $a=e$, $b=f$, $c=g$ and $d=h$. Such pairs of angles are

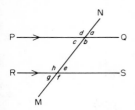

Figure 7.1

called **corresponding angles**.

(iii) $c=e$ and $b=h$. Such pairs of angles are called **alternate angles**.

(iv) $b+e=180°$ and $c+h=180°$. Such pairs of angles are called **interior angles**.

Triangles

6   A triangle is a figure enclosed by three straight lines. The sum of the three angles of a triangle is equal to 180°.

Types of triangles:

7   (i) An **acute-angled triangle** is one in which all the angles are acute, i.e. all the angles are less than 90°.

(ii) A **right-angled triangle** is one which contains a right angle.

(iii) An **obtuse-angled triangle** is one which contains an obtuse angle, i.e. one angle which lies between 90° and 180°.

(iv) An **equilateral triangle** is one in which all the sides and all the angles are equal (i.e. each 60°).

(v) An **isosceles triangle** is one in which two angles and two sides are equal.

(vi) A **scalene triangle** is one with unequal angles and therefore unequal sides.

8   With reference to *Figure 7.2*:

(i) Angles *A, B* and *C* are called **interior angles** of the triangle.

(ii) Angle θ is called an exterior angle of the triangle and is equal to the sum of the two opposite interior angles, i.e. $\theta = A + C$.

(iii) $a + b + c$ is called the **perimeter** of the triangle.

9   With reference to *Figure 7.3*, the side opposite the right angle (side *b*) is called the **hypotenuse**.

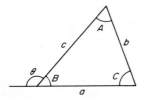

Figure 7.2

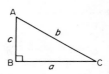

Figure 7.3

The **theorem of Pythagoras** states:
'In any right-angled triangle, the square on the hypotenuse is equal to the sum of the squares on the other two sides.'

Hence $b^2 = a^2 + c^2$.

10   Two triangles are said to be **congruent** if they are equal in all respects, i.e. three angles and three sides in one triangle are equal to three angles and three sides in the other triangle. Two triangles are congruent if:

(i) the three sides of one are equal to the three sides of the other (SSS),

(ii) they have two sides of the one equal to two sides of the other, and if the angles included by three sides are equal (SAS),

(iii) two angles of the one are equal to two angles of the other and any side of the first is equal to the corresponding side of the other (ASA), or

(iv) their hypotenuses are equal and if one other side of one is equal to the corresponding side of the other (RHS).

11   Two triangles are said to be **similar** if the angles of one triangle are equal to the angles of the other triangle. With reference to *Figure 7.4*: triangles ABC and PQR are similar and the corresponding sides are in proportion to each other,

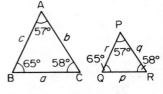

Figure 7.4

i.e. $\dfrac{p}{a} = \dfrac{q}{b} = \dfrac{r}{c}$.

96

Circles

12   A circle is a plain figure enclosed by a curved line, every point on which is equidistant from a point within, called the **centre**.

Properties of circles:

13   (i) The distance from the centre to the curve is called the **radius**, **r**, of the circle (see OP in *Figure 7.5*).

(ii) The boundary of a circle is called the **circumference**, *c*.

(iii) Any straight line passing through the centre and touching the circumference at each end is called the **diameter**, *d*, (see QR in *Figure 7.5*). Thus $d = 2r$.

(iv) The ratio $\dfrac{\text{circumference}}{\text{diameter}}$ = a constant for any circle.

This constant is denoted by the Greek letter π (pronounced 'pie'), where $\pi = 3.14159$, correct to 5 decimal places.

Hence $\dfrac{c}{d} = \pi$ or $c = \pi d$ **or** $c = 2\pi r$.

(v) A **semicircle** is one half of the whole circle.

(vi) A **quadrant** is one quarter of a whole circle.

(vii) A **tangent** to a circle is a straight line which meets the circle in one point only and does not cut the circle when produced. AC in *Figure 7.5* is a tangent to the circle since it touches the curve at point B only. If radius OB is drawn, then angle ABO is a right angle.

(viii) A **sector** of a circle is the part of a circle between radii (for example, the portion OXY of *Figure 7.6* is a sector). If a sector is less than a semicircle it is called a **minor sector**, if greater than a semicircle it is called a **major sector**.

(ix) A **chord** of a circle is any straight line which divides the circle into two parts and is terminated at each end by the circumference. ST, in *Figure 7.6*, is a chord.

(x) A **segment** is the name given to the parts into which a circle is divided by a chord. If the segment is less than a semicircle it is called a **minor segment** (see shaded area in *Figure 7.6*). If the segment is greater than a semicircle it is called a **major segment** (see the unshaded area in *Figure 7.6*).

(xi) An **arc** is a portion of the circumference of a circle.

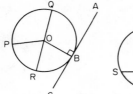

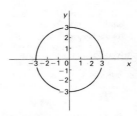

Figure 7.5   Figure 7.6   Figure 7.7

Figure 7.8   Figure 7.9

The distance SRT in *Figure 7.6* is called a **minor arc** and
the distance SXYT is called a **major arc**.

(xii) The angle at the centre of a circle, subtended by an
arc, is double the angle at the circumference subtended by
the same arc. With reference to *Figure 7.7*,

Angle AOC = 2 × angle ABC.

(xiii) The angle in a semicircle is a right angle (see angle
BQP in *Figure 7.7*). One **radian** is defined as the angle
subtended at the centre of a circle by an arc equal in
length to the radius. With reference to *Figure 7.8*, for arc
length l, θ radians $= \dfrac{l}{r}$ or $l = r\theta$, where θ is in radians.

When $l =$ whole circumference $(= 2\pi r)$ then $\theta = \dfrac{l}{r} = \dfrac{2\pi r}{r} = 2\pi$,

i.e. 2π **radians** $= 360°$ **or** π **radians** $= 180°$. Thus

$1 \text{ rad} = \dfrac{180°}{\pi} = 57.30°$, correct to 2 decimal places. Since

π rad $= 180°$, then $\dfrac{\pi}{2}$ rad $= 90°$, $\dfrac{\pi}{3}$ rad $= 60°$, $\dfrac{\pi}{4}$ rad $= 45°$, and so on.

(xiv) The **equation of a circle**, centre at origin, radius r, is given by:

$$x^2 + y^2 = r^2$$

Figure 7.9 shows a circle $x^2 + y^2 = 9$.

8 Graphs

Straight line graphs

1   A **graph** is a pictorial representation of information showing how one quantity varies with another related quantity. The most common method of showing a relationship between two sets of data is to use **cartesian** or **rectangular axes** as shown in *Figure 8.1*.

2   The points on a graph are called **co-ordinates**. Point A in *Figure 8.1* has the co-ordinates (3,2), i.e. 3 units in the *x* direction and 2 units in the *y* direction. Similarly, point B has co-ordinates (−4, 3) and C has co-ordinates (−3, −2). The origin has co-ordinates (0, 0).

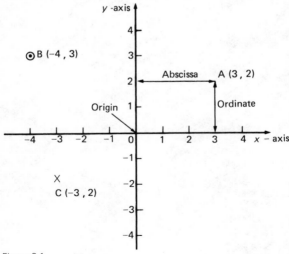

Figure 8.1

3   The horizontal distance of a point from the vertical axis is
called the **abscissa** and the vertical distance from the horizontal
axis is called the **ordinate**.

4   Let a relationship between two variables x and y be $y = 3x + 2$.
When $x = 0$, $y = 3(0) + 2 = 2$,
when $x = 1$, $y = 3(1) + 2 = 5$,
when $x = 2$, $y = 3(2) + 2 = 8$, and so on.

Thus co-ordinates $(0, 2)$, $(1, 5)$ and $(2, 8)$ have been produced
from the equation by selecting arbitrary values of x, and are shown
plotted in *Figure 8.2*. When the points are joined together a
straight-line graph results.

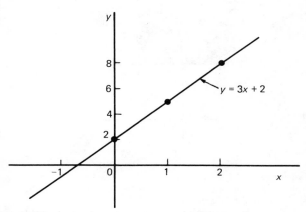

Figure 8.2

5   The **gradient or slope** of a straight line is the ratio of the
change in the value of y to the change in the value of x between
any two points on the line. If, as x increases $(\rightarrow)$, y also increases
$(\uparrow)$, then the gradient is positive.

In *Figure 8.3(a)*

the gradient of AC $= \dfrac{\text{change in } y}{\text{change in } x} = \dfrac{\text{CB}}{\text{BA}} = \dfrac{7-3}{3-1} = \dfrac{4}{2} = 2$.

If as x increases $(\rightarrow)$, y decreases $(\downarrow)$, then the gradient is negative.
In *Figure 8.3(b)*,

the gradient of DF $= \dfrac{\text{change in } y}{\text{change in } x} = \dfrac{\text{FE}}{\text{ED}} = \dfrac{11-2}{-3-0} = \dfrac{9}{-3} = -3$.

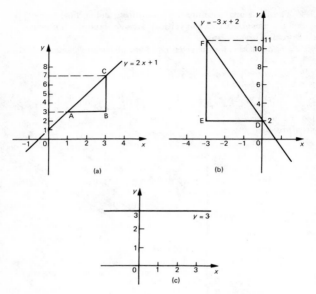

Figure 8.3

Figure 8.3(c) shows a straight line graph $y = 3$. Since the straight line is horizontal the gradient is zero.

6   The value of y when $x = 0$ is called the **y-axis intercept**. In *Figure 8.3(a)* the y-axis intercept is 1 and in *Figure 8.3(b)* is 2.

7   If the equation of a graph is of the form $y = mx + c$, where m and c are constants, the graph will always be a straight line, m representing the gradient and c the y-axis intercept.

    Thus, $y = 5x + 2$ represents a straight line of gradient 5 and y-axis intercept 2. Similarly, $y = -3x - 4$ represents a straight line of gradient -3 and y-axis intercept -4.

8   When a set of co-ordinate values are given or are obtained experimentally and it is believed that they follow a law of the form $y = mx + c$, then if a straight line can be drawn reasonably close to most of the co-ordinate values when plotted, this verifies that a law of the form $y = mx + c$ exists. From the graph, constants m (i.e. gradient) and c (i.e. y-axis intercept) can be determined. This technique is called **'determination of law'**. (See paras 11 to 14.)

9   (a) The process of finding co-ordinate values in between the given information is called **interpolation**.

    (b) The process of finding co-ordinate values which are outside of a given range of values is called **extrapolation**.

Summary of general rules to be applied when drawing graphs

10   (i) Give the graph a title clearly explaining what is being illustrated.

(ii) Choose scales such that the graph occupies as much space as possible on the graph paper being used.

(iii) Choose scales so that interpolation is made as easy as possible. Usually scales such as 1 cm = 1 unit, or 1 cm = 2 units, or 1 cm = 10 units are used. Awkward scales such as 1 cm = 3 units or 1 cm = 7 units should not be used.

(iv) The scales need not start at zero, particularly when starting at zero produces an accumulation of points within a small area of the graph paper.

(v) The co-ordinates, or points, should be clearly marked. This may be done either by a cross, or by a dot and circle, or just by a dot (see *Figure 8.1*).

(vi) A statement should be made next to each axis explaining the numbers represented with their appropriate units.

(vii) Sufficient numbers should be written next to each axis without cramping.

Reduction of non-linear laws to linear form

11   Frequently, the relationship between two variables, say x and y, is not a linear one, i.e. when x is plotted against y a curve results. In such cases the non-linear equation may be modified to the linear form, $y = mx + c$, so that the constants, and thus the law relating the variables can be determined. This technique is called **'determination of law'**.

12   Some examples of the reduction of equations to linear form include:

(i) $y = ax^2 + b$ compares with $Y = mX + c$, where $m = a$, $c = b$ and $X = x^2$. Hence y is plotted vertically against x^2 horizontally to produce a straight line graph of gradient 'a' and y-axis intercept 'b'.

(ii) $y = \dfrac{a}{x} + b$

y is plotted vertically against $\dfrac{1}{x}$ horizontally to produce a straight line graph of gradient 'a' and y-axis intercept 'b'.

(iii) $y = ax^2 + bx$

Dividing both sides by x gives $\dfrac{y}{x} = ax + b$.

Comparing with $Y = mX + c$ shows that $\dfrac{y}{x}$ is plotted vertically against x horizontally to produce a straight line graph of gradient 'a' and $\dfrac{y}{x}$-axis intercept 'b'.

(iv) $y = ax^n$

Taking logarithms to a base of 10 of both sides gives:

$$\lg y = \lg(ax^n) = \lg a + \lg x^n$$

i.e. $\lg y = n \lg x + \lg a$

which compares with $Y = mX + c$,
which shows that $\lg y$ is plotted vertically against $\lg x$ horizontally to produce a straight line graph of gradient n and $\lg y$-axis intercept $\lg a$.

(v) $y = ab^x$

Taking logarithms to a base of 10 of both sides gives:

$$\lg y = \lg (ab^x)$$
i.e. $\lg y = \lg a + \lg b^x$
i.e. $\lg y = x \lg b + \lg a$
or $\lg y = (\lg b)x + \lg a$

which compares with $Y = mX + c$,
which shows that $\lg y$ is plotted vertically against x horizontally to produce a straight line graph of gradient $\lg b$ and $\lg y$-axis intercept $\lg a$.

(vi) $y = ae^{bx}$.

Taking logarithms to a base of e of both sides gives:

$$\ln y = \ln (ae^{bx})$$
i.e. $\ln y = \ln a + \ln e^{bx}$
i.e. $\ln y = \ln a + bx \ln e$
i.e. $\ln y = bx + \ln a$

which compares with $Y = mX + c$,
which shows that $\ln y$ is plotted vertically against x horizontally to produce a straight line graph of gradient b and $\ln y$-axis intercept $\ln a$.

Logarithmic graph paper

13   (i) Graph paper is available where the scale markings
along the horizontal and vertical axes are proportional to
the logarithms of the numbers. Such graph paper is called
log-log graph paper.

(ii) A **logarithmic scale** is shown in *Figure 8.4* where the
distance between, say 1 and 2 is proportional to $\lg 2 - \lg 1$; i.e. 0.3010 of the total distance from 1 to 10. Similarly,
the distance between 7 and 8 is proportional to $\lg 8 - \lg 7$,
i.e. 0.05799 of the total distance from 1 to 10. Thus the
distance between markings progressively decreases as the
numbers increase from 1 to 10.

Figure 8.4

(iii) With log-log graph paper the scale markings are
from 1 to 9, and this pattern can be repeated several times.
The number of times the pattern of markings is repeated
on an axis signifies the number of **cycles**. When the
vertical axis has, say, 3 sets of values from 1 to 9 and the
horizontal axis has 2 sets of values from 1 to 9, then this
log-log graph paper is called 'log 3 cycle × 2 cycle'.
Many different arrangements are available ranging from
'log 1 cycle × 1 cycle' through to 'log 5 cycle × 5 cycle'.

(iv) To depict a set of values, say, from 0.4 to 161 on an
axis of log-log graph paper, 4 cycles are required, from
0.1 to 1, 1 to 10, 10 to 100 and 100 to 1000.

*Plotting graphs of the form $y=ax^n$ in linear form using
logarithmic graph paper*

14   From para 12(iv), $y=ax^n$ reduces to $\lg y = n \lg x + \lg a$,
showing that $\lg y$ is plotted vertically against $\lg x$ horizontally to
produce a straight line graph. With log-log graph paper available
x and y may be plotted directly, without having to firstly determine
their logarithms. The gradient of the straight line gives n and $\lg a$
is given by the point where the line passes through the ordinate $x=1$
(i.e. where $\lg 1 = 0$). A straight line graph representing $y=ax^n$ is
shown on log-log graph paper in *Figure 8.5* corresponding to the
values:

x	0.41	0.63	0.92	1.36	2.17	3.95
y	0.45	1.21	2.89	7.10	20.79	82.46

To evaluate constants a and n, two methods are available.

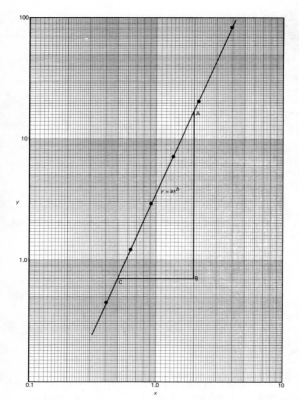

Figure 8.5

Method 1

Any two points on the straight line, say points A and C are selected, and AB and BC are measured.

The gradient, $n = \dfrac{AB}{BC} = \dfrac{11.5 \text{ units}}{5 \text{ units}} = \mathbf{2.3}$

Since $\lg y = n \lg x + \lg a$, when $x = 1$, $\lg x = 0$ and $\lg y = \lg a$. The straight line crosses the ordinate $x = 1.0$ at $y = 3.5$

Hence $\lg a = \lg 3.5$, i.e. $a = 3.5$

Method 2

Any two points on the straight line, say points A and C, are selected. A has co-ordinates (2, 17.25) and C has co-ordinates (0.5, 0.7).

Since $y = ax^n$ then $17.25 = a(2)^n$ (1)

         and $0.7 = a(0.5)^n$ (2)

i.e., two simultaneous equations are produced and may be solved for a and n.

Dividing equation (1) by equation (2) to elimate a gives:

$$\frac{17.25}{0.7} = \frac{(2)^n}{(0.5)^n} = \left(\frac{2}{0.5}\right)^n .$$

i.e., $24.643 = 4^n$

     $\lg 24.643 = n \lg 4$

     and $n = \dfrac{\lg 24.643}{\lg 4} = 2.3$, correct to 2 significant figures.

Substituting $n = 2.3$ in equation (1) gives:

     $17.25 = a(2)^{2.3}$

i.e. $a = \dfrac{17.25}{(2)^{2.3}} = \dfrac{17.25}{4.925} = 3.5$, correct to 2 significant figures.

Hence the law of the graph is $y = 3.5x^{2.3}$

Plotting graphs of the form $y = a.b^n$ in linear form using log-linear graph paper

From para 12(v), $y = a.b^n$ reduces to $\lg y = (\lg b)x + \lg a$, showing that $\lg y$ is plotted vertically against x horizontally to produce a straight line graph. In this case, graph paper having a linear horizontal scale and a logarithmic vertical scale may be used. This type of graph paper is called **log-linear graph paper**, and is specified by the number of cycles on the logarithmic scale. For example, graph paper having 3 cycles on the logarithmic scale is called 'log 3 cycle x linear' graph paper.

     The gradient of the straight line gives $\lg b$ and the intercept on the vertical axis where $x = 0$ gives $\lg a$. If it is not possible to read this directly then two points on the straight line are selected and simultaneous equations solved.

Plotting graphs of the form $y = ae^{kx}$ in linear form using log-linear graph paper

From para 12(vi), $y = ae^{kx}$ reduces to $\ln y = kx + \ln a$, showing that $\ln y$ is plotted vertically against x horizontally to produce a straight line graph. Since $\ln y = 2.3026 \lg y$, i.e., $\ln y = (\text{a constant})(\lg x)$,

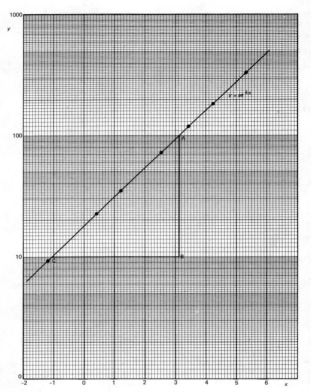

Figure 8.6

the same log–linear graph paper can be used for Naperian logarithms as for logarithms to a base of 10.

A graph of the form $y = ae^{kx}$ is shown in *Figure 8.6*. Gradient of straight line

$$k = \frac{AB}{BC} = \frac{\ln 100 - \ln 10}{3.12 - (-1.08)} = \frac{2.3026}{4.20}$$
$$= 0.55, \text{ correct to 2 significant figures.}$$

Since $\ln y = kx + \ln a$, when $x = 0$, $\ln y = \ln a$, i.e., $y = a$. The vertical axis intercept value at $x=0$ is 18, hence $a=18$. The law of the graph is thus $y=18e^{0.55x}$

When, for example, x is 3.8,

$y=18e^{0.55(3.8)}=18e^{2.09}=18(8.0849)=146$

When, for example, y is 85, $85=18e^{0.55x}$. Hence $e^{0.55x}=\dfrac{85}{18}=4.7222$

and $0.55x=\ln 4.7222=1.5523$. Hence $x=\dfrac{1.5523}{0.55}=2.82$

Graphical solution of equations

15 **Linear simultaneous equations** in two unknowns may be solved graphically by:

      (i) plotting the two straight lines on the same axes, and
      (ii) noting their point of intersection.

The co-ordinates of the point of intersection give the required solution. Thus, for example, to solve graphically the simultaneous equations:

$$2x-y=4$$
$$x+y=5$$

the equations are first rearranged into $y=mx+c$ form, giving:

$$y=2x-4 \qquad (1)$$
$$y=-x+5 \qquad (2)$$

Only three co-ordinates need be calculated for each graph since both are straight lines.

x	0	1	2
$y=2x-4$	-4	-2	0

x	0	1	2
$y=-x+5$	5	4	3

Each of the graphs are plotted as shown in *Figure 8.7*. The point of intersection is at $(3, 2)$ and since this is the only point which lies simultaneously on both lines then $x=3$, $y=2$ is the solution of the simultaneous equations.

16 A general **quadratic equation** is of the form $y=ax^2+bx+c$, where a, b and c are constants and a is not equal to zero. A graph of a quadratic equation always produces a shape called a **parabola**.

17 The gradient of the curve between O and A and between B and C in *Figure 8.8* is positive, whilst the gradient between A and B is negative. Points such as A and B are called **turning points**. At A the gradient is zero and, as x increases, the gradient of the curve

109

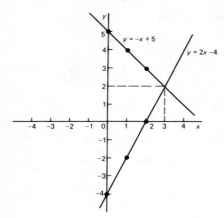

Figure 8.7

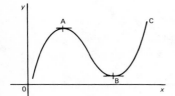

Figure 8.8

changes from positive just before A to negative just after. Such a point is called a **maximum value**. At B the gradient is also zero and, as x increases, the gradient of the curve changes from negative just before B to positive just after. Such a point is called a **minimum value**.

Quadratic graphs

18   (a) $y = ax^2$

Graphs of $y = x^2$, $y = 3x^2$ and $y = \frac{1}{2}x^2$ are shown in *Figure 8.9*.

All have minimum values at the origin (0, 0). Graphs of

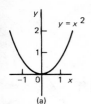
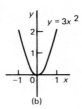
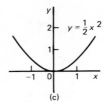

(a)   (b)   (c)

Figure 8.9

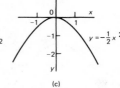

(a)   (b)   (c)

Figure 8.10

$y = -x^2$, $y = -3x^2$ and $y = -\frac{1}{2}x^2$ are shown in *Figure 8.10*.

All have maximum values at the origin $(0, 0)$.

When $y = ax^2$

  (i) curves are symmetrical about the y-axis,
  (ii) the magnitude of 'a' affects the gradient of the curve and
  (iii) the sign of 'a' determines whether it has a maximum or minimum value.

(b) $y = ax^2 + c$

    Graphs of $y = x^2 + 3$, $y = x^2 - 2$, $y = -x^2 + 2$ and $y = -2x^3 - 1$ are shown in *Figure 8.11*.
    When $y = ax^2 + c$:
    (i) curves are symmetrical about the y-axis,
    (ii) the magnitude of 'a' affects the gradient of the curve, and
    (iii) the constant 'c' is the y-axis intercept.

(c) $y = ax^2 + bx + c$

    Whenever 'b' has a value other than zero the curve is

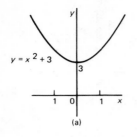

(a)

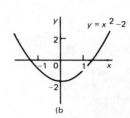

(b

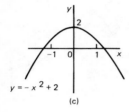

(c)

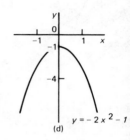

(d)

Figure 8.11

displaced to the right or left of the y-axis. When $\dfrac{b}{a}$ is positive, the curve is displaced $\dfrac{b}{2a}$ to the left of the y-axis, as shown in *Figure 8.12(a)*. When $\dfrac{b}{a}$ is negative the curve is displaced $\dfrac{b}{2a}$ to the right of the y-axis, as shown in *Figure 8.12(b)*

19   **Quadratic equations** of the form $ax^2 + bx + c = 0$ may be solved graphically by:

(i) plotting the graph $y = ax^2 + bx + c$, and
(ii) noting the points of intersection on the x-axis, (i.e. where $y = 0$).

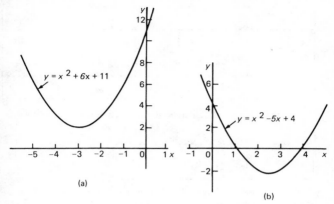

(a)

(b)

Figure 8.12

The x values of the points of intersection give the required solutions since at these points both $y=0$ and $ax^2 + bx + c = 0$. The number of solutions, or roots of a quadratic equation depends on how many times the curve cuts the x-axis and there can be no real roots (as in *Figure 8.12(a)* or one root (as in *Figures 8.9 and 8.10*) or two roots (as in *Figure 8.12(b)*).

From *Figure 8.9(b)*, the solution of the quadratic equation $x^2 - 5x + 4 = 0$ is $x = 1$ or $x = 4$.

20 The solution of **linear and quadratic equations simultaneously** may be achieved graphically by:

(i) plotting the straight line and parabola on the same axes, and
(ii) noting the points of intersection.

The co-ordinates of the points of intersection give the required solutions. Thus, for example, to determine graphically the values of x and y which simultaneously satisfies the equations:

$$y = 2x^2 - 3x - 4 \text{ and}$$
$$y = 2 - 4x$$

a table of values is first drawn up for each equation.
$y = 2x^2 - 3x - 4$ is a parabola and a table of values is drawn up as shown below.

x	-2	-1	0	1	2	3
$2x^2$	8	2	0	2	8	18
$-3x$	6	3	0	-3	-6	-9
-4	-4	-4	-4	-4	-4	-4
y	10	1	-4	-5	-2	5

$y = 2 - 4x$ is a straight line and only three co-ordinates need be calculated.

x	0	1	2
y	2	-2	-6

The two graphs are shown plotted in *Figure 8.13* and the points of intersection, shown as A and B are at co-ordinates $(-2, 10)$ and $(1\frac{1}{2}, -4)$. Hence the simultaneous solutions occur when $x = -2$, $y = 10$ and when $x = 1\frac{1}{2}$, $y = -4$. (These solutions may be checked by substituting into each of the original equations.)

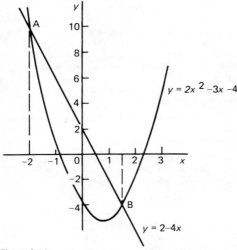

Figure 8.13

21 A **cubic equation** of the form $ax^3 + bx^2 + cx + d = 0$ may be solved graphically by:

      (i) plotting the graph $y = ax^3 + bx^2 + cx + d$, and
      (ii) noting the points of intersection on the x-axis (i.e. where $y = 0$).

The x-values of the points of intersection give the required solution since at these points both $y = 0$ and $ax^3 + bx^2 + cx + d = 0$. The number of solutions, or roots of a cubic equation depends on how many times the curve cuts the x-axis and there can be one, two or three possible roots, as shown in *Figure 8.14.*

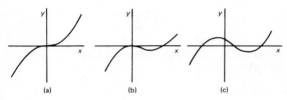

    (a)   (b)   (c)

Figure 8.14

Thus, for example, to solve the cubic equation
$4x^3 - 8x^2 - 15x + 9 = 0$:
Let $y = 4x^3 - 8x^2 - 15x + 9$. A table of values is drawn up as shown below.

x	-2	-1	0	1	2	3
$4x^3$	-32	-4	0	4	32	108
$-8x^2$	-32	-8	0	-8	-32	-72
$-15x$	30	15	0	-15	-30	-45
$+9$	9	9	9	9	9	9
y	-25	12	9	-10	-21	0

A graph of $y = 4x^3 - 8x^2 - 15x + 9$ is shown in *Figure 8.15.* The graph crosses the x-axis (where $y = 0$) at $x = -1\frac{1}{2}$, $x = \frac{1}{2}$ and $x = 3$ and these are the solutions to the cubic equations $4x^3 - 8x^2 - 15x + 9 = 0$.

115

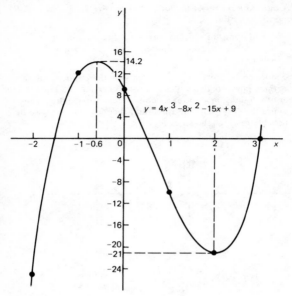

Figure 8.15

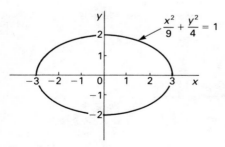

Figure 8.16

116

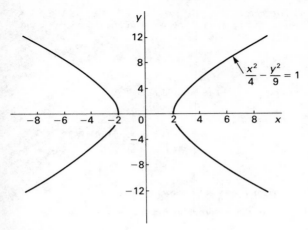

Figure 8.17

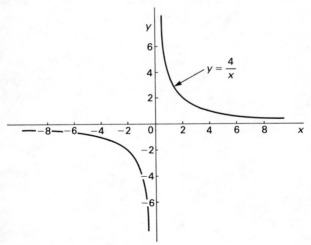

Figure 8.18

117

The ellipse

22   The equation of an ellipse, centre at origin, semi-axes a and b is given by:

$$\frac{x^2}{a^2} + \frac{y^2}{b^2} = 1$$

The ellipse $\dfrac{x^2}{9} + \dfrac{y^2}{4} = 1$ is shown in *Figure 8.16.*

The hyperbola

23   The equation of a hyperbola is of the form:

$$\frac{x^2}{a^2} - \frac{y^2}{b^2} = 1$$

The hyperbola $\dfrac{x^2}{4} - \dfrac{y^2}{9} = 1$ is shown in *Figure 8.17.*

The equation of a **rectangular hyperbola** is of the form:

$$y = \frac{a}{x}$$

The rectangular hyperbola $y = \dfrac{4}{x}$ is shown in *Figure 8.18.*

9 Mensuration

1   Mensuration is a branch of mathematics concerned with the determination of lengths, areas and volumes.

2   A **polygon** is a closed plane figure bounded by straight lines. A polygon which has:

     (i) 3 sides is called a **triangle**
     (ii) 4 sides is called a **quadrilateral**
    (iii) 5 sides is called a **pentagon**
    (iv) 6 sides is called a **hexagon**
     (v) 7 sides is called a **heptagon**
    (vi) 8 sides is called a **octagon**

3   There are five types of **quadrilateral**, these being:

(i) rectangle, (ii) square, (iii) parallelogram, (iv) rhombus, (v) trapezium. (The properties of these are given in paragraphs 4 to 8.)

If the opposite corners of any quadrilateral are joined by a straight line, two triangles are produced. Since the sum of the angles of a triangle is 180°, the sum of the angles of a quadrilateral is 360°.

4   In a **rectangle**, shown in *Figure 9.1*:

    (i) all four angles are right angles,
    (ii) opposite sides are parallel and equal in length, and
    (iii) diagonals AC and BD are equal in length and bisect one another.

5   In a **square**, shown in *Figure 9.2*:

    (i) all four angles are right angles,
    (ii) opposite sides are parallel,
    (iii) all four sides are equal in length, and
    (iv) diagonals PR and QS are equal in length and bisect one another at right angles.

6   In a **parallelogram**, shown in *Figure 9.3*:

    (i) opposite angles are equal,
    (ii) opposite sides are parallel and equal in length, and
    (iii) diagonals WY and XZ bisect one another.

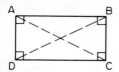

Figure 9.1

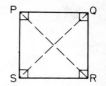

Figure 9.2

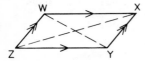

Figure 9.3

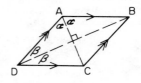

Figure 9.4

7 In a **rhombus**, shown in *Figure 9.4*:

(i) opposite angles are equal,
(ii) opposite angles are bisected by a diagonal,
(iii) opposite sides are parallel,
(iv) all four sides are equal in length, and
(v) diagonals AC and BD disect one another at right angles.

8 In a **trapezium**, shown in *Figure 9.5*:

(i) only one pair of sides is parallel.

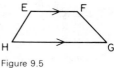

Figure 9.5

9 **Areas of plane figures** (see *Table 9.1*).
10 **Volumes and surface areas of regular solids** (see *Table 9.2*).
11 (i) The **frustrum of a pyramid or cone** is the portion remaining when a part containing the vertex is cut off by a plane parallel to the base. (See *Table 9.2(v)*).
(ii) The **volume of a frustum of a pyramid or cone** is given by the volume of the whole pyramid or cone minus the volume of the small pyramid or cone cut off.

120

Table 9.1

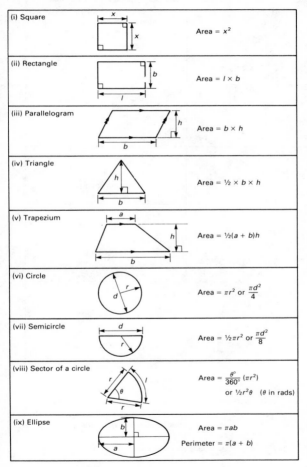

(i) Square	Area $= x^2$
(ii) Rectangle	Area $= l \times b$
(iii) Parallelogram	Area $= b \times h$
(iv) Triangle	Area $= \frac{1}{2} \times b \times h$
(v) Trapezium	Area $= \frac{1}{2}(a + b)h$
(vi) Circle	Area $= \pi r^2$ or $\frac{\pi d^2}{4}$
(vii) Semicircle	Area $= \frac{1}{2}\pi r^2$ or $\frac{\pi d^2}{8}$
(viii) Sector of a circle	Area $= \frac{\theta°}{360°}(\pi r^2)$ or $\frac{1}{2}r^2\theta$ (θ in rads)
(ix) Ellipse	Area $= \pi ab$ Perimeter $= \pi(a + b)$

Table 9.2

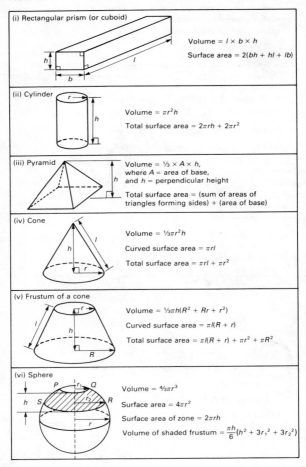

(i) Rectangular prism (or cuboid)

Volume $= l \times b \times h$

Surface area $= 2(bh + hl + lb)$

(ii) Cylinder

Volume $= \pi r^2 h$

Total surface area $= 2\pi rh + 2\pi r^2$

(iii) Pyramid

Volume $= \frac{1}{3} \times A \times h$,
where A = area of base,
and h = perpendicular height

Total surface area = (sum of areas of
triangles forming sides) + (area of base)

(iv) Cone

Volume $= \frac{1}{3}\pi r^2 h$

Curved surface area $= \pi rl$

Total surface area $= \pi rl + \pi r^2$

(v) Frustum of a cone

Volume $= \frac{1}{3}\pi h(R^2 + Rr + r^2)$

Curved surface area $= \pi l(R + r)$

Total surface area $= \pi l(R + r) + \pi r^2 + \pi R^2$

(vi) Sphere

Volume $= \frac{4}{3}\pi r^3$

Surface area $= 4\pi r^2$

Surface area of zone $= 2\pi rh$

Volume of shaded frustum $= \frac{\pi h}{6}(h^2 + 3r_1^2 + 3r_2^2)$

(iii) The **surface area of the sides of a frustum of a pyramid or cone** is given by the surface area of the whole pyramid or cone minus the surface area of the small pyramid or cone cut off. This gives the lateral surface area of the frustum. If the total surface area of the frustum is required then the surface area of the two parallel ends are added to the lateral surface area.

(iv) A **frustum of a sphere** is the portion contained between two parallel planes. In *Table 9.2(vi)*, PQRS is a frustum of the sphere.

(v) A **zone of a sphere** is the curved surface of a frustum.

Areas of irregular figures

12   Areas of irregular plane surfaces may be approximately determined by using (a) a planimeter, (b) the trapezoidal rule, (c) the mid-ordinate rule, and (d) Simpson's rule. Such methods may be used, for example, by engineers estimating areas of indicator diagrams of steam engines, surveyors estimating areas of plots of land or naval architects estimating areas of water planes or transverse sections of ships.

(a) A **planimeter** is an instrument for directly measuring small areas bounded by an irregular curve.

(b) **Trapezoidal rule**. To determine the area PQRS in *Figure 9.6*:

(i) Divide base PS into any number of equal intervals, each of width d, (the greater number of intervals, the greater the accuracy).

(ii) Accurately measure ordinates y_1, y_2, y_3, etc.

(iii) Area PQRS $= d\left[\dfrac{y_1 + y_7}{2} + y_2 + y_3 + y_4 + y_5 + y_6\right]$

In general, the trapezoidal rule states:

$$\textbf{Area} = \begin{pmatrix}\textbf{width of}\\\textbf{interval}\end{pmatrix}\left[\frac{1}{2}\begin{pmatrix}\textbf{first} +\\\textbf{last ordinate}\end{pmatrix} + \begin{array}{c}\textbf{sum of}\\\textbf{remaining ordinates}\end{array}\right]$$

(c) **Mid-ordinate rule**. To determine the area ABCD of *Figure 9.7*:

(i) Divide base AD into any number of equal intervals, each of width d, (the greater the number of intervals, the greater the accuracy).

(ii) Erect ordinates in the middle of each interval (shown by broken lines in *Figure 9.7*).

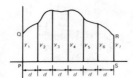

Figure 9.6

Figure 9.7

(iii) Accurately measure ordinates y_1, y_2, y_3, etc.

(iv) Area ABCD $= d(y_1 + y_2 + y_3 + y_4 + y_5 + y_6 + y_7)$

In general, the mid-ordinate rule states:

Area = (width of interval)(sum of mid-ordinates)

(d) **Simpson's rule** To determine the area PQRS of *Figure 9.6*:

(i) Divide base PS into an even number of intervals, each of width d, (the greater the number of intervals, the greater the accuracy).

(ii) Accurately measure ordinates y_1, y_2, y_3, etc.

(iii) Area PQRS $= \dfrac{d}{3}[(y_1 + y_7) + 4(y_2 + y_4 + y_6) + 2(y_3 + y_5)]$

In general, Simpson's rule states:

$$\text{Area} = \frac{1}{3}\binom{\text{width of}}{\text{interval}}\left[\binom{\text{first} +}{\text{last ordinate}} + 4\binom{\text{sum of}}{\text{even ordinates}} + 2\binom{\text{sum of remaining}}{\text{odd ordinates}}\right]$$

Volume of irregular solids

13   If the cross-sectional areas A_1, A_2, A_3, etc., of an irregular solid bounded by two parallel planes are known at equal intervals of width d (as shown in *Figure 9.8*), then by Simpson's rule:

$$\textbf{Volume, } V = \frac{d}{3}[(A_1 + A_7) + 4(A_2 + A_4 + A_6) + 2(A_3 + A_5)]$$

Mean or average value of a waveform

14   (i) The mean or average value, y, or the waveform shown in *Figure 9.9* is given by:

$$y = \frac{\text{area under curve}}{\text{length of base, } b}$$

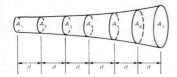

Figure 9.8

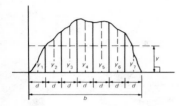

Figure 9.9

(ii) If the mid-ordinate rule is used to find the area under the curve,

$$\text{then}: y = \frac{\text{sum of mid-ordinates}}{\text{number of mid-ordinates}}$$

$$\left(=\frac{y_1 + y_2 + y_3 + y_4 + y_5 + y_6 + y_7}{7} \text{ for } \textit{Figure 9.9}\right)$$

(iii) For a **sine-wave**, the mean or average value:
  (a) over one complete cycle is zero (see *Figure 9.10(a)*).
  (n) over half a cycle is 0.637 × maximum value, or
    $\dfrac{2}{\pi}$ × maximum value,
  (c) of a full-wave rectified waveform (see *Figure 9.10(b)*) is
    0.637 × maximum value,
  (d) of a half-wave rectified waveform (see *Figure 9.10(c)*) is
    0.318 × maximum value, or $\dfrac{1}{\pi}$ × maximum value.

Prismoidal rule for finding volumes

15   The prismoidal rule applies to a solid of length x divided by only three equidistant plane areas, A_1, A_2 and A_3 as shown in *Figure 9.11* and is merely an extension of Simpsons rule for volumes (see para 13).

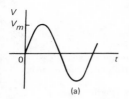

(a)

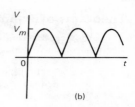

(b)

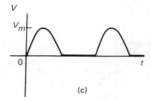

(c)

Figure 9.10

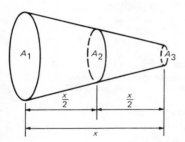

Figure 9.11

With reference to *Figure 9.11*, Volume,

$$\textbf{Volume, } V = \frac{x}{6}[A_1 + 4A_2 + A_3]$$

The prismoidal rule gives precise values of volume for regular solids such as pyramids, cones, spheres and prismoids.

Theorems of Pappus

16   (i) The **first theorem of Pappus** states:

'If a plane curve is rotated about an axis in its own plane but not intersecting it, the surface area generated is given by the product of the arc length and the distance moved by the centroid of the curve.

i.e. Surface area generated = arc length × distance moved through by the centroid.

In *Figure 9.12*, if the curve of length s is rotated one revolution about the x-axis and C is the position of the centroid of the curve

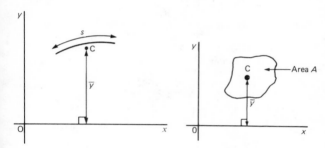

Figure 9.12   Figure 9.13

then the distance moved by the centroid when the curve revolves one revolution about OX is $2\pi\bar{y}$ (i.e., the circumference of a circle).

Hence by Pappus' theorem, **surface area** = arc length × $2\pi\bar{y}$

$$= 2\pi s\bar{y} \text{ square units.}$$

(ii) The **second theorem of Pappus** states:

'If a plane area is rotated about an axis in its own plane but not intersecting it, the volume of the solid formed is given by the product of the area and the distance moved by the centroid of the area.'

i.e. Volume generated = area × distance moved through by the centroid.

In *Figure 9.13*, let C be the centroid of area A, and let $\bar{y}$ be the perpendicular distance of C from axis OX, then the distance moved by the centroid when area A makes one revolution about OX is $2\pi\bar{y}$ (i.e. the circumference of a circle).

Hence by Pappus' theorem, Volume generated = area × $2\pi\bar{y}$
i.e. **$V = 2\pi A\bar{y}$ cubic units.**

Centroids of simple shapes

17   A **lamina** is a thin, flat sheet having uniform thickness. The **centre of gravity** of a lamina is the point where it balances perfectly, i.e. the lamina's **centre of mass**. When dealing with a shape or area (i.e. a lamina of negligible thickness and mass) the term **centre of area** or **centroid** is used for the point where the centre of gravity of a lamina of that shape would lie.

18   (i) The centroid C of a **rectangle** lies on the intersection of the diagonals (see *Figure 9.14(a)*).

       (ii) The centroid C of a **triangle** lies on the intersection of its medians, a median being a line which joins the vertices of a triangle with the mid-point of the opposite side. It may be shown that the centroid lies at one-third of the perpendicular height above any side as base (see *Figure 9.14(b)*).

       (iii) The centroid C of a **circle** lies at its centre (see *Figure 9.14(c)*).

       (iv) The centroid C of a **semicircle** of radius r lies on the centre line at a distance $\dfrac{4r}{3\pi}$ from the diameter (see *Figure 9.14(d)*).

19   The **first moment of area** is defined as the product of the area and the perpendicular distance of its centroid from a given

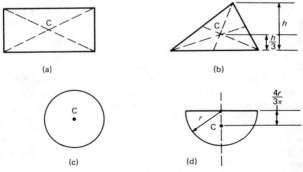

(a)   (b)

(c)   (d)

Figure 9.14

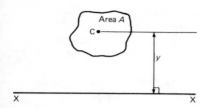

Figure 9.15

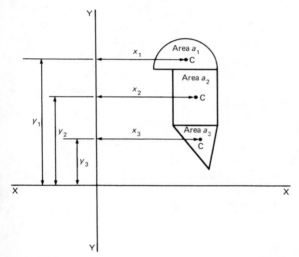

Figure 9.16

axis in the plane of the area. In *Figure 9.15*, the first moment of area A about axis XX is given by (Ay) cubic units.

20   A **composite area** consists of two or more areas having different shapes joined together. The centroid of a composite area is found by dividing the whole area into parts, the centroids of which are known, and then taking moments (i.e. finding the first moment of area) about two orthogonal axes (i.e. two axes lying in the same plane and at right angles to each other). For the composite area shown in *Figure 9.16*:

Sum of moments about YY, $\Sigma ax = a_1x_1 + a_2x_2 + a_3x_3$
Sum of moments about XX, $\Sigma ay = a_1y_1 + a_2y_2 + a_3y_3$

If $A = a_1 + a_2 + a_3$ and $\bar{x}$ and $\bar{y}$ are the distances of the centroid of the composite area about axes YY and XX respectively, then:

$$A\bar{x} = \Sigma ax, \text{ from which } \bar{x} = \frac{\Sigma ax}{A} = \frac{\text{first moment of area about YY}}{\text{total area}}$$

and

$$A\bar{y} = \Sigma ay, \text{ from which } \bar{y} = \frac{\Sigma ay}{A} = \frac{\text{first moment of area about XX}}{\text{total area}}$$

Thus, for example, to find the centroid of the metal template shown in *Figure 9.17* about AB and CD (the circular area being removed) a tabular approach is used as shown below.

Part	Area a cm^2	Distance of centroid from AB (i.e. x cm)	First moment of area about AB (i.e. ax cm^3)	Distance of centroid from CD (i.e. y cm)	First moment or area about CD (i.e. ay cm^3)
Triangle	$\frac{1}{2}(8.0)(12.0)$ $= 48.0$	$\frac{1}{3}(8.0)$ $= 2.667$	$(48.0)(2.667)$ $= 128.0$	$20.0 + \frac{1}{3}(12.0)$ $= 24.0$	$(48.0)(24.0)$ $= 1152.0$
Rectangle	$(20.0)(8.0)$ $= 160.0$	$\frac{1}{2}(8.0)$ $= 4.0$	$(160.0)(4.0)$ $= 640.0$	$\frac{1}{2}(20.0)$ $= 10.0$	$(160.0)(10.0)$ $= 1600.0$
Circle	$-\pi(2.0)^2$ $= -12.566$ (minus since circle is removed)	3.0	$(-12.566)(3.0)$ $= -37.70$	5.0	$(-12.566)(5.0)$ $= -62.83$
Semicircle	$\frac{1}{2}\pi(4.0)^2$ $= 25.133$	4.0	$(25.133)(4.0)$ $= 100.5$	$\frac{-4(4.0)}{3\pi}$ $= -1.698$ (minus since) below CD)	$(25.133)(-1.698)$ $= -42.68$
$\Sigma a = A = 220.567$		$\Sigma ax = 830.8$		$\Sigma ay = 2646.49$	

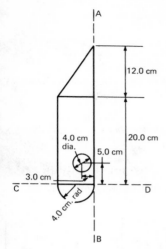

Figure 9.17

If $\bar{x}$ and $\bar{y}$ are the distances of the centroid from AB and CD respectively, then,

$$A\bar{x} = \Sigma ax, \text{ from which } \bar{x} = \frac{\Sigma ax}{A} = \frac{830.8}{220.567} = 3.77 \text{ cm},$$

and

$$A\bar{y} = \Sigma ay, \text{ from which } \bar{y} = \frac{\Sigma zy}{A} = \frac{2646.49}{220.567} = 12.0 \text{ cm}$$

Hence the centroid lies at a point 3.77 cm to the left of AB and 12.0 cm above CD.

10 **Trigonometry**

1   Trigonometry is concerned with the measurements of the sides and angles of triangles and their relationship with each other.

Trigonometric ratios of acute angles

2   With reference to the right angled triangle shown in *Figure 10.1*:

(i)   sine $\theta = \dfrac{\text{opposite side}}{\text{hypotenuse}}$,   i.e. $\sin \theta = \dfrac{b}{c}$

(ii) cosine $\theta = \dfrac{\text{adjacent side}}{\text{hypotenuse}}$,   i.e. $\cos \theta = \dfrac{a}{c}$

(iii) tangent $\theta = \dfrac{\text{opposite side}}{\text{adjacent side}}$,   i.e. $\tan \theta = \dfrac{b}{c}$

(iv) secant $\theta = \dfrac{\text{hypotenuse}}{\text{adjacent side}}$,   i.e. $\sec \theta = \dfrac{c}{a}$

(v)   cosectant $\theta = \dfrac{\text{hypotenuse}}{\text{opposite side}}$, i.e. $\operatorname{cosec} \theta = \dfrac{c}{b}$

(vi) cotangent $\theta = \dfrac{\text{adjacent side}}{\text{opposite side}}$, i.e. $\cot \theta = \dfrac{a}{b}$

Identities

3   A trigonometric identity is an expression that is true for all values of the unknown variable. (The sign '≡' means 'is identical to'.)

4   From para 2,

(i) $\dfrac{\sin \theta}{\cos \theta} = \dfrac{\frac{b}{c}}{\frac{a}{c}} = \dfrac{b}{a} = \tan \theta$, i.e. $\tan \theta = \dfrac{\sin \theta}{\cos \theta}$

(ii) $\dfrac{\cos\theta}{\sin\theta}=\dfrac{\dfrac{a}{c}}{\dfrac{b}{c}}=\dfrac{a}{b}=\cot\theta$, i.e., $\cot\theta=\dfrac{\cos\theta}{\sin\theta}$

(iii) $\sec\theta=\dfrac{1}{\cos\theta}$

(iv) $\operatorname{cosec}\theta=\dfrac{1}{\sin\theta}$

(v) $\cot\theta=\dfrac{1}{\tan\theta}$

Secants, cosecants and cotangents are called the **reciprocal ratios**.

5   Applying Pythagoras' theorem to the right-angled triangle shown in *Figure 10.1* gives:

$$a^2+b^2=c^2 \tag{1}$$

Dividing each term of equation (1) by c^2 gives:

$$\dfrac{a^2}{c^2}+\dfrac{b^2}{c^2}=\dfrac{c^2}{c^2}$$

i.e. $\left(\dfrac{a}{c}\right)^2+\left(\dfrac{b}{c}\right)^2=1$

$(\cos\theta)^2+(\sin\theta)^2=1$

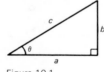

Figure 10.1

Hence

$$\cos^2\theta+\sin^2\theta\equiv1 \tag{2}$$

Dividing each term of equation (1) by a^2 gives:

$$\dfrac{a^2}{a^2}+\dfrac{b^2}{a^2}=\dfrac{c^2}{a^2}$$

i.e. $1+\left(\dfrac{b}{a}\right)^2=\left(\dfrac{c}{a}\right)^2$

Hence

$$1+\tan^2\theta\equiv\sec^2\theta \tag{3}$$

Dividing each term of equation (1) by b^2 gives:

$$\dfrac{a^2}{b^2}+\dfrac{b^2}{b^2}=\dfrac{c^2}{b^2}$$

i.e. $\left(\dfrac{a}{b}\right)^2 + 1 = \left(\dfrac{c}{b}\right)^2$

Hence

$$\cot^2\theta + 1 \equiv \csc^2\theta \tag{4}$$

Fractional and surd forms of trigonometric ratios

6   (i) In *Figure 10.2* ABC is an equilateral triangle of side 2 units. AD bisects angle A and bisects the side BC. Using Pythagoras' theorem on triangle ABD gives: $AD = \sqrt{(2^2 - 1^2)} = \sqrt{3}$. Hence

$$\sin 30^\circ = \frac{BD}{AB} = \frac{1}{2}; \quad \cos 30^\circ = \frac{AD}{AB} = \frac{\sqrt{3}}{2}; \quad \tan 30^\circ = \frac{BD}{AD} = \frac{1}{\sqrt{3}}$$

$$\sin 60^\circ = \frac{AD}{AB} = \frac{\sqrt{3}}{2}; \quad \cos 60^\circ = \frac{BD}{AB} = \frac{1}{2}; \quad \tan 60^\circ = \frac{AD}{BD} = \sqrt{3}$$

(ii) In *Figure 10.3*, PQR is an isosceles triangle with PQ=QR=1 unit. By Pythagoras' theorem,
$PR = \sqrt{(1^2 + 1^2)} = \sqrt{2}$

Hence $\sin 45^\circ = \dfrac{1}{\sqrt{2}}$; $\cos 45^\circ = \dfrac{1}{\sqrt{2}}$; $\tan 45^\circ = 1$

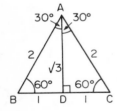

Figure 10.2

Figure 10.3

(iii) A quantity which is not exactly expressible as a rational number is called a **surd**. For example, $\sqrt{2}$ and $\sqrt{3}$ are called surds because they cannot be expressed as a fraction and the decimal part may be continued indefinitely. For example, $\sqrt{2} = 1.4142136\ldots$

(iv) From paras. (i) and (ii), sin 30° = cos 60°, sin 45° = cos 45° and sin 60° = cos 30°. In general, sin θ = cos(90° − θ) and cosθ = sin(90° − θ). For example, it may be checked from tables that sin 25° = cos 65°, sin 42° = cos 48°, cos 84° 10' = sin 5° 50', and so on.

Angles of any magnitude

(i) *Figure 10.4* shows rectangular axes XX' and YY' intersecting at origin 0. As with graphical work, measurements made to the right and above 0 are positive whilst those to the left and downwards are negative. Let OA be free to rotate about 0. By convention, when OA moves anticlockwise angular measurement is considered positive, and vice versa.

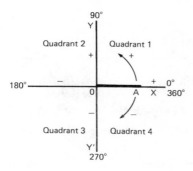

Figure 10.4

(ii) Let OA be rotated anticlockwise so that θ_1 is any angle in the first quadrant and let perpendicular AB be constructed to form the right-angled triangle OAB (see *Figure 10.5*). Since all three sides of the triangle are positive, all six trigonometric ratios are positive in the first quadrant. (Note: OA is always positive since it is the radius of a circle.)

(iii) Let OA be further rotated so that θ_2 is any angle in the second quadrant and let AC be constructed to form the right-angled triangle OAC. Then:

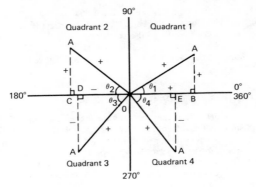

Figure 10.5

$$\sin\theta_2 = \frac{+}{+} = +, \quad \cos\theta_2 = \frac{-}{+} = -, \quad \tan\theta_2 = \frac{+}{-} = -,$$

$$\operatorname{cosec}\theta_2 = \frac{+}{+} = +, \quad \sec\theta_2 = \frac{+}{-} = -, \quad \cot\theta_2 = \frac{-}{+} = -.$$

(iv) Let OA be further rotated so that θ_2 is any angle in the third quadrant and let AD be constructed to form the right-angled triangle OAD. Then: $\sin\theta_3 = \frac{-}{+} = -$ (and hence $\operatorname{cosec}\theta_3$ is $-$), $\cos\theta_3 = \frac{-}{+} = -$ (and hence $\sec\theta_3$ is $-$), $\tan\theta_3 = \frac{-}{-} = +$ (and hence $\cot\theta_3$ is $+$).

(v) Let OA be further rotated so that θ_4 is any angle in the fourth quadrant and let AE be constructed to form the right-angled triangle OAE. Then: $\sin\theta_4 = \frac{-}{+} = -$ (and hence $\operatorname{cosec}\theta_4$ is $-$), $\cos\theta_4 = \frac{+}{+} = +$ (and hence $\sec\theta_4$ is $+$), $\tan\theta_4 = \frac{-}{+} = -$ (and hence $\cot\theta_4$ is $-$).

(vi) The results obtained in (ii) to (v) are summarised in *Figure 10.6*. The letters underlined spell the word CAST when starting in the fourth quadrant and moving in an anticlockwise direction.

(vii) To evaluate $\sin 240°$ using tables:

(a) Sketch rectangular axes and mark on angles and the word CAST as shown in *Figure 10.7*.

136

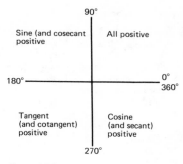

Figure 10.6

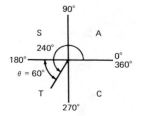

Figure 10.7

(b) Mark on the sketch an angle of 240°.
(c) Determine angle θ $(=240°-180°=60°)$, which is always measured to the horizontal.
(d) Use tables to evaluate sin 60° (=0.8660).
(e) Since T is shown in the third quadrant, then only tangent and contangent are positive in this quadrant, i.e. sine is negative. Hence sin 240°= - sin 60°= -0.8660.

(viii) A knowledge of angles of any magnitude is necessary even when using a calculator. For example, when finding the angles between 0° and 360° whose tangent is 1.7629 by calculator (i.e., inverse tangent 1.7629) only one answer is normally given, i.e. 60.436°. However, there is another angle whose tangent is + 1.7629, this being in in the third quadrant, i.e. 180°+ 60.436°= 240.436°.

The solution of triangles and their areas

8   (i) To **'solve a triangle'** means 'to find the values of
unknown sides and angles'. If a triangle is **right-angled**,
trigonometric ratios and the theorem of Pythagoras may
be used for its solution. However, for a **non-right-angled
triangle**, trigonometric ratios and Pythagoras' theorem
cannot be used. Instead, two rules, called the sine rule
and the cosine rule are used.

(ii) With reference to
triangle ABC of *Figure 10.8*:
the sine rule states:

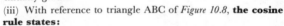

$$\frac{a}{\sin A} = \frac{b}{\sin B} = \frac{c}{\sin C}$$

The rule may be used only
when:
(a) 1 side and any 2 angles
are initially given, and
(b) 2 sides and an angle
(not the included angle) are
initially given.

Figure 10.8

(iii) With reference to triangle ABC of *Figure 10.8*, **the cosine
rule states:**

$$a^2 = b^2 + c^2 - 2bc \cos A$$
$$\text{or } b^2 = a^2 + c^2 - 2ac \cos B$$
$$\text{or } c^2 = a^2 + b^2 - 2ab \cos C$$

The rule may be used only when:
(a) 2 sides and the included angle are initially given; or
(b) 3 sides are initially given.

(iv) **The area of any triangle** such as ABC as *Figure 10.8* is
given by:

(a) $\frac{1}{2} \times$ base $\times$ perpendicular height; or

(b) $\frac{1}{2}ab \sin C$ or $\frac{1}{2}ac \sin B$ or $\frac{1}{2}bc \sin A$; or

(b) $\sqrt{[s(s-a)(s-b)(s-c)]}$, where $s = \frac{a+b+c}{2}$

Lengths and areas on an inclined plane

9   In *Figure 10.9*, rectangle ADEF is a plane inclined at an

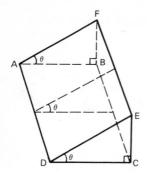

Figure 10.9

angle of θ to the horizontal plane ABCD

$\cos \theta = \dfrac{DC}{DE}$, from which, $DE = \dfrac{DC}{\cos \theta}$

Hence the line of greatest slope on an inclined plane is given by:

$\left(\dfrac{1}{\cos \theta} \right)$ (its projection on to the horizontal plane)

Area of ADEF $= (AD)(DE) = (AD)\left(\dfrac{DC}{\cos \theta} \right)$

$$= \left(\dfrac{1}{\cos \theta} \right) \textbf{(area of horizontal plane)}$$

10 **The angle between a line and a plane** is defined as the angle between the line and its projection on the plane. In *Figure 10.10*, the line PQ meets a plane at P. If QR is constructed perpendicular to the plane then the projection of PQ on the plane is PR. The angle between the line PQ and the plane in *Figure 10.10* is θ.

11 *Figure 10.11* shows two planes ABCD and ABEF intersecting along the line AB. **The angle between the planes** is defined as the angle between any two straight lines drawn on each plane which meet at, and are perpendicular to, the line of intersection of the planes. In *Figure 10.11*, PQ and PR are both perpendicular to AB thus the angle between the two intersecting planes is $\angle$ RPQ.

12 Three-dimensional triangulation problems rely on the ability to (i) visualise the problem and (ii) solve triangles. A clearly labelled sketch is thus usually invaluable. Determining the location,

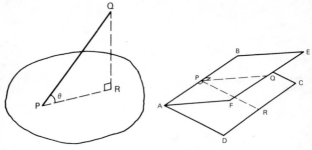

Figure 10.10   Figure 10.11

speed and direction of moving objects, such as ships and aircraft, usually involves the solution of three-dimensional problems.

13   (i) If, in *Figure 10.12(a)*, BC represents horizontal ground and AB a vertical flagpole, then **the angle of elevation** of the top of the flagpole, A, from the point C is the angle that the imaginary straight line AC must be raised (or elevated) from the horizontal CB, i.e., angle θ.

(ii) If, in *Figure 10.12(b)*, PQ represents a vertical cliff and R a ship at sea, then **the angle of depression** of the ship from point P is the angle through which the imaginary straight line PR must be lowered (or depressed) from the horizontal to the ship, i.e., angle ϕ. (Note, $\angle$ PRQ is also ϕ – alternate angles between parallel lines.)

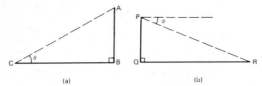

(a)   (b)

Figure 10.12

14   **Bearings** provide a method of specifying the position of an object with respect to the points of the compass. There are two methods of stating bearings:

140

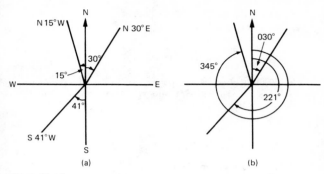

Figure 10.13

(i) A bearing of N 30° E means an angle of 30° measured from the north towards east as shown in *Figure 13(a)*. Similarly, bearings of S 41° W and N 15° W are also shown.

(ii) The angle denoting the bearing is measured from north in a clockwise direction, north being considered as 0°. Three figures are always stated. Bearings of 030°, 221° and 345° are shown in *Figure 10.13(b)*, and are equivalent to N 30° E, S 41° W and N 15° W respectively.

Graphs of trigonometric functions

15   (i) In *Figure 10.14*, OR represents a vector that is free to rotate anticlockwise about 0 at a velocity of ω rad/s. A rotating vector is called a phasor. After a time *t* seconds

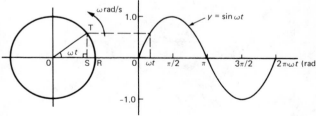

Figure 10.14

OR will have turned through an angle ωt radians (shown as angle TOR in *Figure 10.14*). If ST is constructed perpendicular to OR, then

$$\sin \omega t = \frac{ST}{OT}, \text{ i.e. } ST = OT \sin \omega t.$$

If all such vertical components are projected on to a graph of y against ωt, a **sine wave** results of maximum value OR.

(ii) If phasor OR of *Figure 10.14* makes one revolution (i.e. 2π radians) in T sec, then the angular velocity, $\omega = \dfrac{2\pi}{T}$ rad/s, from which $T = \dfrac{2\pi}{\omega}$ sec. T is known as the **periodic time**.

(iii) The number of complete cycles occurring per second is called the **frequency** f.

$$\text{Frequency} = \frac{\text{number of cycles}}{\text{second}} = \frac{1}{T} = \frac{\omega}{2\pi} \text{ Hz, i.e.,}$$

$$f = \frac{\omega}{2\pi} \text{ Hz.}$$

Hence angular velocity, $\omega = 2\pi f$ rad/s.

16
(i) *Figure 10.15* shows graph of $y = \sin \omega t$, for $\omega = \dfrac{1}{2}$, 1, 2 and 3 rad/s. The graphs are plotted either from a calculated table of values or by the rotating vector approach.

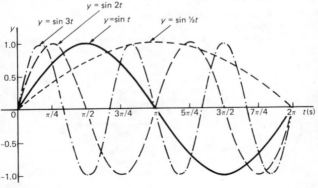

Figure 10.15

142

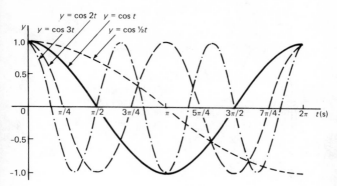

Figure 10.16

(ii) *Figure 10.16* shows graphs of $y = \cos \omega t$, for $\omega = \dfrac{1}{2}$, 1, 2 and 3 rad/s.

(iii) Each of the waveforms shown in *Figures 10.15 and 10.16* repeat themselves after period time T $\left(= \dfrac{2\pi}{\omega} \text{ seconds}\right)$ and such functions are known as **periodic functions**. It is noted from the graphs that in 2 π sec:

$\sin \dfrac{1}{2}t$ and $\cos \dfrac{1}{2}t$ complete $\dfrac{1}{2}$ cycle,

$\sin t$ and $\cos t$ complete 1 cycle,
$\sin 2t$ and $\cos 2t$ complete 2 cycles, and
$\sin 3t$ and $\cos 3t$ complete 3 cycles.

17 (i) *Figure 10.17* shows graphs of $y = \sin^2 t$ and $y = \cos^2 t$ which may be obtained by drawing up a table of values. Both are periodic functions of periodic time π seconds and both contain only positive values.
(ii) *Figure 10.18(a)* shows a graph of $y = \sin^2 3t$, which has a periodic time of $\dfrac{\pi}{3}$ seconds.

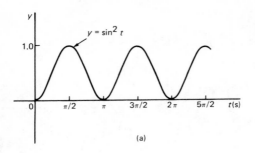

(a)

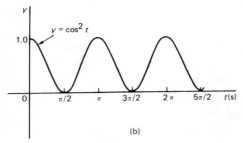

(b)

Figure 10.17

(iii) *Figure 10.18(b)* shows a graph of $y = \cos^2 2t$, which has a periodic time of $\dfrac{\pi}{2}$ seconds.

(iv) In general, if $y = \sin^2 \omega t$ or $y = \cos^2 \omega t$ then the periodic time is given by $\dfrac{\pi}{\omega}$ seconds.

(v) Graphs of the form $y = \sin^2 \omega t$ and $y = \cos^2 \omega t$ are **not** sine waves and **cannot** be produced by the rotating vector approach.

The general form of a sine wave, $R \sin(\omega t \pm \alpha)$

18   **Amplitude** is the name given to the maximum or peak value of a sine wave. Each of the graphs shown in *Figures 10.15 and 10.16*

144

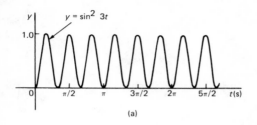

(a)

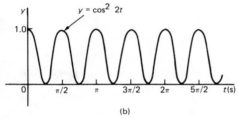

(b)

Figure 10.18

have an amplitude of 1. However, if $y = 4 \sin \omega t$, the maximum value, and thus amplitude is 4.

19 *Lagging and leading angles*

(i) A sine or cosine curve may not always start at zero. To show this a periodic function is represented by $y = \sin(\omega t \pm \alpha)$ or $y = \cos(\omega t \pm \alpha)$, where α is a phase displacement compared with $y = \sin \omega t$ or $y = \cos \omega t$.

(ii) By drawing up a table of values, a graph of $y = \sin\left(\omega t - \dfrac{\pi}{3}\right)$ may be plotted as shown in *Figure 10.19*. If $y = \sin \omega t$ is assumed to start at zero, then $y = \sin\left(\omega t - \dfrac{\pi}{3}\right)$ starts $\dfrac{\pi}{3}$ radians later $\left(\text{i.e., has a zero value } \dfrac{\pi}{3} \text{ rads later}\right)$. Thus $y = \sin\left(\omega t - \dfrac{\pi}{3}\right)$ is said to **lag** $y = \sin \omega t$ by $\dfrac{\pi}{3}$ rads.

(iii) By drawing up a table of values, a graph of

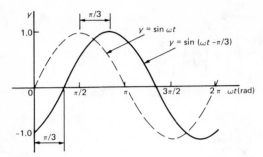

Figure 10.19

$y = \cos\left(\omega t + \dfrac{\pi}{4}\right)$ may be plotted as shown in *Figure 10.20*. If $y = \cos \omega t$ is assumed to start at zero, then $y = \cos\left(\omega t + \dfrac{\pi}{4}\right)$ starts $\dfrac{\pi}{4}$ radians earlier $\left(\text{i.e. has a value of } 1, \dfrac{\pi}{4} \text{ rad earlier}\right)$. Thus $y = \cos\left(\omega t + \dfrac{\pi}{4}\right)$ is said to **lead** $y = \cos \omega t$ by $\dfrac{\pi}{4}$ rad.

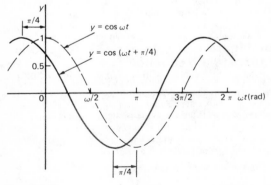

Figure 10.20

(iv) Generally, a graph of $y = \sin(\omega t - \alpha)$ lags $y = \sin \omega t$ by angle α, and a graph of $y = \sin(\omega t + \alpha)$ leads $y = \sin \omega t$ by angle α.

(v) A cosine curve is the same shape as a sine curve but starts $\dfrac{\pi}{2}$ rad earlier, i.e. leads by $\dfrac{\pi}{2}$ rad. Hence $\cos \omega t = \sin\left(\omega t + \dfrac{\pi}{2}\right)$.

(vi) In terms of time, $y = R \sin(\omega t - \alpha)$ lags $y = R \sin \omega t$ by $\dfrac{\alpha}{\omega}$ sec, and $y = R \sin(\omega t + \alpha)$ leads $y = R \sin \omega t$ by $\dfrac{\alpha}{\omega}$ sec.

Summary

20   Given a general sinusoidal periodic function $y = R \sin(\omega t \pm \alpha)$, then: R = amplitude, ω = angular velocity, $\dfrac{2\pi}{\omega}$ = periodic time, T, $\dfrac{\omega}{2\pi}$ = frequency, f, and α = angle of lead or lag (compared with $y = R \sin \omega t$).

Thus, for example, an alternating voltage $v = 300 \sin(200\pi t + 0.26)$ volts represents a sine wave of amplitude (i.e., maximum value) 300 V, angular velocity $\omega = 200\pi$ rad/s, periodic time $T = \dfrac{2\pi}{\omega} = \dfrac{2\pi}{200\pi} = 0.01$ s or 10 ms, frequency $f = \dfrac{1}{T} = 100$ Hz and the phase angle is 0.26 rad or $14°\,54'$ leading $300 \sin 200\pi t$.

Combination of two periodic functions of the same frequency

21   There are a number of instances in engineering and science where waveforms combine and where it is required to determine the single phasor (called the resultant) which could replace two or more separate phasors. Uses are found in electrical alternating current theory, in mechanical vibrations, in the addition of forces and with sound waves. There are several methods of determining the resultant and two such methods are shown below.

(i) *Plotting the periodic functions graphically* This may be achieved by sketching the separate functions on the same axes and then adding (or subtracting) ordinates at regular intervals.

(Alternatively, a table of values may be drawn up before plotting the resultant waveform.)

Thus, for example, $y_1 = 4 \sin \omega t$ and $y_2 = 3 \sin\left(\omega t - \dfrac{\pi}{3}\right)$ are shown plotted in *Figure 10.21*. Ordinates are added at 15° intervals and the resultant is shown by the broken line. The amplitude of the resultant is 6.1 and it lags y_1 by 25° or 0.436 rads. Hence the sinusoidal expression for the resultant waveform is:

$$y_R = 6.1 \sin(\omega t - 0.436).$$

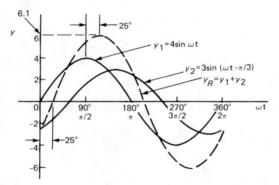

Figure 10.21

(ii) Resolution of phasors by drawing or calculation

The resultant of two periodic functions may be found from their relative positions when the time is zero. For example, $y_1 = 4 \sin \omega t$ and $y_2 = 3 \sin\left(\omega t - \dfrac{\pi}{3}\right)$ then each may be represented as phasors as shown in *Figure 10.22*, y_1 being 4 units long and drawn horizontally and y_2 being 3 units long, lagging y_1 by $\dfrac{\pi}{3}$ radians or 60°.

To determine the resultant of $y_1 + y_2$, y_1 is drawn horizontally as shown in *Figure 10.23*, and y_2 is joined to the end of y_1 at 60° to the horizontal. The resultant is given by y_R. This is the same as the diagonal of a parallelogram which is shown completed in *Figure 10.24*.

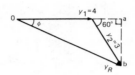

Figure 10.22   Figure 10.23

Resultant, y_R in *Figures 10.23 and 10.24* is determined either by
(a) scaled drawing and measurement, or
(b) by use of the cosine rule, (and then sine rule to calculate angle ϕ), or
(c) by determining horizontal and vertical components of lengths oa and ab in *Figure 10.23*, and then using Pythagoras' theorem to calculate ob.

In this case, by calculation, $y_R = 6.083$ and angle $\phi = 25.28°$ or 0.441 rads. Thus the resultant may be expressed in sinusoidal form as $y_R = 6.083 \sin(\omega t - 0.441)$. If the resultant phasor, $y_R = y_1 - y_2$ is required, then y_2 is still 3 units long but is drawn in the opposite direction, as shown in *Figure 10.25*, and y_R is determined by measurement or calculation.

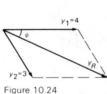

Figure 10.24

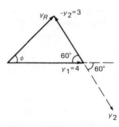

Figure 10.25

Compound angles

22   Angles such as $(A+B)$ or $(A-B)$ are called compound angles since they are the sum or difference of two angles, A and B.
23   The **compound angle formulae** for sines and cosines of the sum and difference of two angles A and B are:

$$\sin(A+B) = \sin A \cos B + \cos A \sin B$$
$$\sin(A-B) = \sin A \cos B - \cos A \sin B$$

149

$$\cos(A + B) = \cos A \cos B - \sin A \sin B$$
$$\cos(A - B) = \cos A \cos B + \sin A \sin B$$

(Note, $\sin(A + B)$ is **not** equal to $(\sin A + \sin B)$, and so on.)

24   The formulae stated in para 23 may be used to derive two further compound-angle formulae:

$$\tan(A + B) = \frac{\tan A + \tan B}{1 - \tan A \tan B}$$
$$\tan(A - B) = \frac{\tan A - \tan B}{1 + \tan A \tan B}$$

25   The compound-angle formulae are true for all values of A and B, and by substituting values of A and B into the formulae they may be shown to be true.

26   (i) $R \sin(\omega t + \alpha)$ represents a sine wave of maximum value R, periodic time $\dfrac{2\pi}{\omega}$, frequency $\dfrac{\omega}{2\pi}$ and leading $R \sin \omega t$ by angle α. (See para 20.)

(ii) $R \sin(\omega t + \alpha)$ may be expanded using the compound-angle formula for $\sin(A + B)$, where $A = \omega t$ and $B = \alpha$.

Hence $R \sin(\omega t + \alpha) = R[\sin \omega t \cos \alpha + \cos \omega t \sin \alpha]$
$$= R \sin \omega t \cos \alpha + R \cos \omega t \sin \alpha$$
$$= (R \cos \alpha) \sin \omega t + (R \sin \alpha) \cos \omega t$$

(iii) If $a = R \cos \alpha$ and $b = R \sin \alpha$, where a and b are constants, then $R \sin(\omega t + \alpha) = a \sin \omega t + b \cos \omega t$. i.e. a sine and cosine function of the same frequency when added produce a sine wave of the same frequency.

(iv) Since $a = R \cos \alpha$, then $\cos \alpha = \dfrac{a}{R}$, and since

$b = R \sin \alpha$, then $\sin \alpha = \dfrac{b}{R}$.

If the values of a and b are known then the values of R and α may be calculated. The relationship between constants a, b, R and α are shown in *Figure 10.26*. From *Figure 10.26* by Pythagoras' theorem

$$R = \sqrt{(a^2 + b^2)}$$

and from trigonometric ratios:

$$\alpha = \arctan \frac{b}{a}.$$

Thus, for example, to express $4.6 \sin \omega t - 7.3 \cos \omega t$ in the form $R \sin(\omega t + \alpha)$.

let $4.6 \sin \omega t - 7.3 \cos \omega t = R \sin(\omega t + \alpha)$
then $4.6 \sin \omega t - 7.3 \cos \omega t = R[\sin \omega t \cos \alpha + \cos \omega t \sin \alpha]$
$$= (R \cos \alpha) \sin \omega t + (R \sin \alpha) \cos \omega t$$

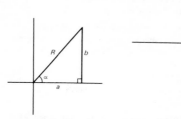

Figure 10.26

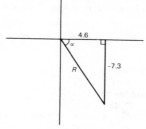

Figure 10.27

Equating coefficients of sin ωt gives: $4.6 = R \cos \alpha$, from which,

$$\cos \alpha = \frac{4.6}{R}$$

Equating coefficients of cos ωt gives $-7.3 = R \sin \alpha$, from which

$$\sin \alpha = \frac{-7.3}{R}$$

There is only one quadrant where cosine is positive and sine is negative, i.e., the fourth quadrant, as shown in *Figure 10.27*. By Pythagoras' theorem: $R = \sqrt{[(4.6)^2 + (-7.3)^2]} = 8.628$. From trigonometric ratios:

$$\alpha = \arctan\left(\frac{-7.3}{4.6}\right) = -57.78° \text{ or } -1.008 \text{ radians.}$$

Hence 4.6 sin $\omega t - 7.3$ cos $\omega t = 8.628$ sin $\omega t - 1.008$).

Double angles

27   (i) If, in the compound-angle formula for $\sin(A + B)$, we let $B = A$ then **sin $2A = 2$ sin A cos A.** Also, for example,

$$\sin 4A = 2 \sin 2 \, A \cos 2 \, A$$

and

$$\sin 8A = 2 \sin 4A \cos 4A, \text{ and so on.}$$

(ii) If, in the compound-angle formula for $\cos(A + B)$, we let $B = A$ then **cos $2A = \cos^2 A - \sin^2 A$**
Since $\cos^2 A + \sin^2 A = 1$, then $\cos^2 A = 1 - \sin^2 A$, and $\sin^2 A = 1 - \cos^2 A$, and two further formula for cos $2A$ can be produced.

151

Thus $\cos 2A = \cos^2 A - \sin^2 A = (1 - \sin^2 A) - \sin^2 A$

i.e., $\quad \mathbf{\cos 2A = 1 - 2 \sin^2 A}$

and $\quad \cos 2A = \cos^2 A - \sin^2 A = \cos^2 A - (1 - \cos^2 A)$

i.e., $\quad \mathbf{\cos 2A = 2 \cos^2 A - 1.}$

Also, for example,

$\cos 4A = \cos^2 2A - \sin^2 2A$ or $1 - 2 \sin^2 2A$ or $2 \cos^2 2A - 1$

and

$\cos 6A = \cos^2 3A - \sin^2 3A$ or $1 - 2 \sin^2 3A$ or $2 \cos^2 3A - 1$,

and so on.

(iii) If, in the compound-angle formula for $\tan(A + B)$, we let $B = A$ then

$$\mathbf{\tan 2A = \frac{2 \tan A}{1 - tan^2 A}}$$

Also, for example, $\tan 4A = \dfrac{2 \tan 2A}{1 - \tan^2 2A}$ and

$$\tan 5A = \frac{2 \tan \frac{5}{2} A}{1 - \tan^2 \frac{5}{2} A} \text{ and so on.}$$

Changing products of sines and cosines into sums or differences

28   (i) $\sin(A + B) + \sin(A - B) = 2 \sin A \cos B$ (from the formulae in para 23)

$$\text{i.e. } \mathbf{\sin A \cos B = \frac{1}{2}\big[\sin(A + B) + \sin(A - B)\big]} \qquad (1)$$

(ii) $\sin(A + B) - \sin(A - B) = 2 \cos A \sin B$

$$\text{i.e. } \mathbf{\cos A \sin B = \frac{1}{2}\big[\sin(A + B) - \sin(A - B)\big]} \qquad (2)$$

(iii) $\cos(A + B) + \cos(A - B) = 2 \cos A \cos B$

$$\text{i.e. } \mathbf{\cos A \cos B = \frac{1}{2}\big[\cos(A + B) + \cos(A - B)\big]} \qquad (3)$$

(iv) $\cos(A + B) - \cos(A - B) = -2 \sin A \sin B$

$$\text{i.e. } \mathbf{\sin A \sin B = -\frac{1}{2}\big[\cos(A + B) - \cos(A - B)\big]} \qquad (4)$$

Thus, for example, $\sin 4x \cos 3x = \dfrac{1}{2}[\sin(4x+3x)+\sin(4x-3x)]$ from equation (1)

$$=\dfrac{1}{2}(\sin 7x + \sin x)$$

and $3 \cos 4t \cos t = 3\left\{\dfrac{1}{2}[\cos(4t+t)+\cos(4t-t)]\right\}$ from equation (3)

$$=\dfrac{3}{2}(\cos 5t + \cos 3t).$$

Changing sums or differences of sines and cosines into products

29   In the compound-angle formulae let $(A+B)=X$ and $(A-B)=Y$. Solving the simultaneous equations gives $A=\dfrac{X+Y}{2}$ and $B=\dfrac{X-Y}{2}$.

Thus $\sin(A+B)+\sin(A-B)=2 \sin A \cos B$. becomes

$$\sin X + \sin Y = 2 \sin\left(\dfrac{X+Y}{2}\right) \cos\left(\dfrac{X-Y}{2}\right) \qquad (5)$$

Similarly,

$$\sin X - \sin Y = 2 \cos\left(\dfrac{X+Y}{2}\right) \sin\left(\dfrac{X-Y}{2}\right) \qquad (6)$$

$$\cos X + \cos Y = 2 \cos\left(\dfrac{X+Y}{2}\right) \cos\left(\dfrac{X-Y}{2}\right) \qquad (7)$$

$$\cos X - \cos Y = -2 \sin\left(\dfrac{X+Y}{2}\right) \sin\left(\dfrac{X-Y}{2}\right) \qquad (8)$$

Thus, for example, $\sin 7x - \sin 3x = 2 \cos\left(\dfrac{7x+3x}{2}\right) \sin\left(\dfrac{7x-3x}{2}\right)$
from equation (6)

$$= 2 \cos 5x \sin 2x$$

and

$$\cos 5\theta - \cos 2\theta = -2 \sin\left(\dfrac{5\theta+2\theta}{2}\right) \sin\left(\dfrac{5\theta-2\theta}{2}\right)$$

from equation (8)

$$= -2 \sin\dfrac{7\theta}{2} \sin\dfrac{3\theta}{2}.$$

11 Hyperbolic functions

1   Functions which are associated with the geometry of the conic section called a hyperbola are called hyperbolic functions. Applications of hyperbolic functions include transmission line theory and catenary problems.

2   By definition:

(i) Hyperbolic sine of x, **sinh** $x = \dfrac{e^x - e^{-x}}{2}$ (1)

'sinh x' is often abbreviated to 'sh x' and is pronounced as 'shine x'.

(ii) Hyperbolic cosine of x, **cosh** $x = \dfrac{e^x + e^{-x}}{2}$ (2)

'cosh x' is often abbreviated to 'ch x' and is pronounced as 'kosh x'.

(iii) Hyperbolic tangent of x, **tanh** $x = \dfrac{\sinh x}{\cosh x} = \dfrac{e^x - e^{-x}}{e^x + e^{-x}}$
(3)

'tanh x' is often abbreviated to 'th x' and is pronounced as 'than x'.

(iv) Hyperbolic cosecant of x, **cosech** $x = \dfrac{1}{\sinh x} = \dfrac{2}{e^x - e^{-x}}$

'cosech x' is pronounced as 'coshec x'. (4)

(v) Hyperbolic secant of x, **sech** $x = \dfrac{1}{\cosh x} = \dfrac{2}{e^x + e^{-x}}$ (5)

'sech x' is pronounced as 'shec x'.

(vi) Hyperbolic cotangent of x, **coth** $x = \dfrac{1}{\tanh x} = \dfrac{e^x + e^{-x}}{e^x - e^{-x}}$
(6)

'coth x' is pronounced as 'koth x'.

Some properties of hyperbolic functions

3   (i) Replacing x by 0 in equation (1) gives:

$$\sinh 0 = \frac{e^0 - e^{-0}}{2} = \frac{1-1}{2} = 0$$

(ii) Replacing x by 0 in equation (2) gives:

$$\cosh 0 = \frac{e^0 + e^{-0}}{2} = \frac{1+1}{2} = 1$$

(iii) If a function of x, $f(-x) = -f(x)$ then $f(x)$ is called an **odd function** of x. Replacing x by $-x$ in equation (1) gives:

$$\sinh(-x) = \frac{e^{-x} + e^{-(-x)}}{2} = \frac{e^{-x} - e^x}{2} = -\left(\frac{e^x - e^{-x}}{2}\right)$$
$$= -\sinh x$$

Replacing x by $-x$ in equation (3) gives:

$$\tanh(-x) = \frac{e^{-x} - e^{-(-x)}}{e^{-x} + e^{-(-x)}} = \frac{e^{-x} - e^x}{e^{-x} + e^x} = -\left(\frac{e^x - e^{-x}}{e^x + e^{-x}}\right)$$
$$= -\tanh x$$

Hence $\sinh x$ and $\tanh x$ are both odd functions, (see para 5), as also are cosech $x\left(=\dfrac{1}{\sinh x}\right)$ and coth $x\left(=\dfrac{1}{\tanh x}\right)$.

(iv) If a function of x, $f(-x) = f(x)$, then $f(x)$ is called an **even function** of x. Replacing x by $-x$ in equation (2) gives:

$$\cosh(-x) = \frac{e^{-x} + e^{-(-x)}}{2} = \frac{e^{-x} + e^x}{2} = \cosh x.$$

Hence $\cosh x$ is an even function, (see para 5), as also is sech $x\left(=\dfrac{1}{\cosh x}\right)$.

4   Tables of exponential and hyperbolic functions are available where values of $\sinh x$ and $\cosh x$ may be read directly, usually from an argument of $x=0$ to $x=6.0$. Values of hyperbolic functions for values of x greater than 6.0 and for values of x of greater accuracy than 2 significant figures are evaluated using the definitions of para 2 or by using a calculator containing hyperbolic functions. For example, $\sinh 0.27 = 0.2733$, $\cosh 0.60 = 1.1855$, $\sinh 3.9 = 24.69$ and $\cosh 6.329 = 280.3$, each correct to 4 significant figures.

Graphs of hyperbolic functions

5   (i) A graph of $y = \sinh x$ may be plotted using values from tables of hyperbolic functions, or by using a calculator. The curve is shown in *Figure 11.1*. Since the graph is symmetrical about the origin, $\sinh x$ is an odd function (as stated in para 2(iii)).

(ii) A graph of $y = \cosh x$ may be plotted using values from tables of hyperbolic functions, or by using a calculator. The curve is shown in *Figure 11.2*. Since the graph is symmetrical about the y-axis, $\cosh x$ is an even function (as stated in para 2(iv)).

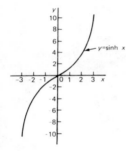

Figure 11.1

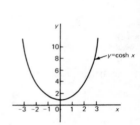

Figure 11.2

The shape of $y = \cosh x$ is that of a heavy rope or chain hanging freely under gravity and is called a **catenary**. Examples include, transmission lines, a telegraph wire or a fisherman's line.

(iii) Graphs of $y = \tanh x$, $y = \operatorname{cosech} x$, $y = \operatorname{sech} x$ and $y = \coth x$ are shown in *Figure 11.3*, values geing determined from the definitions in para 2(iii)–(vi).

Hyperbolic identities

6   For every trigonometric identity there is a corresponding hyperbolic identity. **Hyperbolic identities** may be proved by either (i) replacing sh x by $\left(\dfrac{e^x - e^{-x}}{2}\right)$ and ch x by $\left(\dfrac{e^x + e^{-x}}{2}\right)$, or

(ii) by using **Osborne's rule**, which states:
'*the six trigonometric ratios used in trigonometrical identities relating general*

156

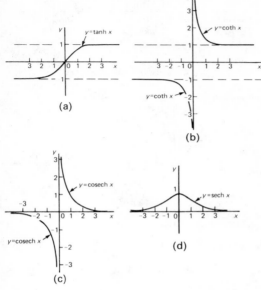

Figure 11.3

angles may be replaced by their corresponding hyperbolic functions, but the sign of any direct or implied product of two sines must be changed'.

For example, since $\cos^2 x + \sin^2 x = 1$ then, by Osborne's rule, $\text{ch}^2 x - \text{sh}^2 x = 1$, i.e. the trigonometric functions have been changed to their corresponding hyperbolic functions and since $\sin^2 x$ is a product of two sines the sign is changed from $+$ to $-$.

Table 11.1 shows some trigonometrical identities and their corresponding hyperbolic identities.

Differentiation of hyperbolic functions

7
(i) $\dfrac{d}{dx}(\sinh x) = \dfrac{d}{dx}\left(\dfrac{e^x - e^{-x}}{2}\right) = \left(\dfrac{e^x - (-e^{-x})}{2}\right) = \left(\dfrac{e^x + e^{-x}}{2}\right)$

$\qquad\qquad = \cosh x$

If $y = \sinh ax$, where 'a' is a constant, then $\dfrac{dy}{dx} = a \cosh ax$.

Table 11.1

Trigonometric identity	Corresponding hyperbolic identity
$\cos^2 x + \sin^2 x = 1$	$\text{ch}^2 x - \text{sh}^2 x = 1$
$1 + \tan^2 x = \sec^2 x$	$1 - \text{th}^2 x = \text{sech}^2 x$
$\cot^2 x + 1 = \csc^2 x$	$\coth^2 x - 1 = \text{cosech}^2 x$
Compound angle formulae	
$\sin(A \pm B) = \sin A \cos B \pm$ $\cos A \sin B$	$\text{sh } (A + B) = \text{sh } A \text{ ch } B \pm$ $\text{ch } A \text{ ch } B$
$\cos(A \pm B) = \cos A \cos B \mp$ $\sin A \sin B$	$\text{ch } (A + B) = \text{ch } A \text{ ch } B \pm$ $\text{sh } A \text{ sh } B$
$\tan (A \pm B) = \dfrac{\tan A \pm \tan B}{1 \mp \tan A \ \tan B}$	$\text{th } (A \pm B) = \dfrac{\text{th } A \pm \text{th } B}{1 \pm \text{th } A \text{ th } B}$
Double angles	
$\sin 2x = 2 \sin x \cos x$	$\text{sh } 2x = 2 \text{ sh } x \text{ ch } x$
$\cos 2x = \cos^2 x - \sin^2 x$	$\text{ch } 2x = \text{ch}^2 x + \text{sh}^2 x$
$\quad = 2 \cos^2 x - 1$	$\quad = 2 \text{ ch}^2 x - 1$
$\quad = 1 - 2 \sin^2 x$	$\quad = 1 + 2 \text{ sh}^2 x$

(ii) $\dfrac{d}{dx}(\cosh x) = \dfrac{d}{dx}\left(\dfrac{e^x + e^{-x}}{2}\right) = \left(\dfrac{e^x + (-e^{-x})}{2}\right)$

$$= \left(\dfrac{e^x - e^{-x}}{2}\right) = \sinh x$$

If $y = \cosh ax$, where 'a' is a constant, then $\dfrac{dy}{dx} = a \sinh ax$.

(iii) Using the quotient rule of differentiation the derivatives of $\tanh x$, $\operatorname{sech} x$, $\operatorname{cosech} x$ and $\coth x$ may be determined using the results of (i) and (ii) and are summarised in the tables of derivatives on page 180.

8   **Equations of the form $a \text{ ch } x + b \text{ sh } x = c$, where a, b and c are constants may be solved either by:**
    (a) plotting graphs of $y = a \text{ ch } x + b \text{ sh } x$ and $y = c$ and noting the points of intersection, or more accurately,
    (b) by adopting the following procedure:

    (i) Change sh x to $\left(\dfrac{e^x - e^{-x}}{2}\right)$ and ch x to $\left(\dfrac{e^x + e^{-x}}{2}\right)$
    (ii) Rearrange the equation into the form $pe^x + qe^{-x} + r = 0$, where p, q and r are constants.

(iii) Multiply each term by e^x, which produces an equation of the form $p(e^x)^2 + re^x + q = 0$ (since $(e^{-x})(e^x) = e^0 = 1$).

(iv) Solve the quadratic equation $p(e^x)^2 + re^x + q = 0$ for e^x by factorising or by using the quadratic formula.

(v) Given $e^x = a$ constant, (obtained by solving the equation in (iv)), take Naperian logarithms of both sides to give $x = \ln$ (constant).

Thus, for example, to solve the equation 2.6 ch $x + 5.1$ sh $x = 8.73$, correct to 4 decimal places, the above procedure is followed.

(i) 2.6 ch $x + 5.1$ sh $x = 8.73$

i.e., $2.6\left(\dfrac{e^x + e^{-x}}{2}\right) + 5.1\left(\dfrac{e^x - e^{-x}}{2}\right) = 8.73$

(ii) $1.3e^x + 1.3e^{-x} + 2.55e^x - 2.55e^{-x} = 8.73$

i.e., $3.85e^x - 1.25e^{-x} - 8.73 = 0$

(iii) $3.85(e^x)^2 - 8.73e^x - 1.25 = 0$

(iv) $e^x = \dfrac{-(-8.73) \pm \sqrt{[(-8.73)^2 - 4(3.85)(-1.25)]}}{2(3.85)}$

$= \dfrac{8.73 \pm \sqrt{95.463}}{7.70} = \dfrac{8.73 \pm 9.7705}{7.70}$

Hence $e^x = 2.4027$ or $e^x = -0.1351$

(v) $x = \ln 2.4027$ or $x = \ln(-0.1351)$ which has no real solution. Hence $x = \mathbf{0.8766}$, correct to 4 decimal places.

Series expansions for cosh x and sinh x

9 (i) By definition

$$e^x = 1 + x + \frac{x^2}{2!} + \frac{x^3}{3!} + \frac{x^4}{4!} + \frac{x^5}{5!} + \dots$$

Replacing x by $-x$ gives:

$$e^{-x} = 1 - x + \frac{x^2}{2!} - \frac{x^3}{3!} + \frac{x^4}{4!} - \frac{x^5}{5!} + \dots$$

(ii) $\cosh x = \dfrac{1}{2}(e^x + e^{-x}) = \dfrac{1}{2}\left[\left(1 + x + \dfrac{x^2}{2!} + \dfrac{x^3}{3!} + \dfrac{x^4}{4!} + \dfrac{x^5}{5!} + \dots\right)\right.$

$\left. + \left(1 - x + \dfrac{x^2}{2!} - \dfrac{x^3}{3!} + \dfrac{x^4}{4!} - \dfrac{x^5}{5!} + \dots\right)\right]$

$= \dfrac{1}{2}\left(2 + \dfrac{2x^2}{2!} + \dfrac{2x^4}{4!} + \dots\right)$

i.e. $\cosh x = 1 + \dfrac{x^2}{2!} + \dfrac{x^4}{4!} + \ldots$ (valid for all values of x)

$\cosh x$ is an even function and contains only even powers of x in its series expansion.

(iii) $\sinh x = \dfrac{1}{2}(e^x - e^{-x}) = \dfrac{1}{2}\left[\left(1 + x + \dfrac{x^2}{2!} + \dfrac{x^3}{3!} + \dfrac{x^4}{4!} + \dfrac{x^5}{5!} + \ldots\right)\right.$

$$\left. - \left(1 - x + \dfrac{x^2}{2!} - \dfrac{x^3}{3!} + \dfrac{x^4}{4!} - \dfrac{x^5}{5!} + \ldots\right)\right]$$

$$= \dfrac{1}{2}\left(2x + \dfrac{2x^3}{3!} + \dfrac{2x^5}{5!} + \ldots\right)$$

i.e. **$\sinh x = x + \dfrac{x^3}{3!} + \dfrac{x^5}{5!} + \cdots$** (valid for all values of x)

$\sinh x$ is an odd function and contains only odd powers of x in its series expansion.

Relationship between trigonometric and hyperbolic functions

10   On page 89, it is shown that

$$\cos \theta + j \sin \theta = e^{j\theta} \tag{7}$$

and

$$\cos \theta - j \sin \theta = e^{-j\theta} \tag{8}$$

Adding equations (7) and (8) gives:

$$\boldsymbol{\cos \theta = \dfrac{1}{2}(e^{j\theta} + e^{-j\theta})} \tag{9}$$

Subtracting equation (8) from equation (7) gives:

$$\boldsymbol{\sin \theta = \dfrac{1}{2j}(e^{j\theta} - e^{-j\theta})} \tag{10}$$

11   Substituting $j\theta$ for θ in equations (9) and (10) gives:

$$\cos j\theta = \dfrac{1}{2}(e^{j(j\theta)} + e^{-j(j\theta)}) \text{ and}$$

$$\sin j\theta = \dfrac{1}{2j}(e^{j(j\theta)} - e^{-j(j\theta)})$$

Since $j^2 = -1$, $\cos j\theta = \dfrac{1}{2}(e^{-\theta} + e^{\theta}) = \dfrac{1}{2}(e^{\theta} + e^{-\theta})$

Hence from para 2, **$\cos j\theta = \cosh \theta$** $\tag{11}$

160

Similarly, $\sin j\theta = \dfrac{1}{2j}(e^{-\theta} - e^{\theta}) = -\dfrac{1}{2j}(e^{\theta} - e^{-\theta})$

$$= \dfrac{-1}{j}\left[\dfrac{1}{2}(e^{\theta} - e^{-\theta})\right] = -\dfrac{1}{j}\sinh\theta \text{ (see para 2).}$$

But $-\dfrac{1}{j} = -\dfrac{1}{j} \times \dfrac{j}{j} = -\dfrac{j}{j^2} = j$, hence **$\sin j\theta = j \sinh\theta$** (12)

Equations (11) and (12) may be used to verify that in all standard trigonometric identities, $j\theta$ may be written for θ and the identity still remains true.

12   From para 2, $\cosh\theta = \dfrac{1}{2}(e^{\theta} + e^{-\theta})$ and substituting $j\theta$ for θ

gives:

$$\cosh j\theta = \dfrac{1}{2}(e^{j\theta} + e^{-j\theta}) = \cos\theta, \text{ from equation (9),}$$

i.e. **$\cosh j\theta = \cos\theta$** (13)

Similarly, from para 2, $\sinh\theta = \dfrac{1}{2}(e^{\theta} - e^{-\theta})$

Substituting $j\theta$ for θ gives:

$$\sinh j\theta = \dfrac{1}{2}(e^{j\theta} - e^{-j\theta}) = j\sin\theta, \text{ from equation (10).}$$

Hence **$\sinh j\theta = j \sin\theta$** (14)

13    $\tan j\theta = \dfrac{\sin j\theta}{\cos j\theta}$

From equations (11) and (12), $\dfrac{\sin j\theta}{\cos j\theta} = \dfrac{j\sinh\theta}{\cosh\theta} = j\tanh\theta$

Hence **$\tan j\theta = j \tanh\theta$** (15)

Similarly, $\tanh j\theta = \dfrac{\sinh j\theta}{\cosh j\theta}$

From equation (13) and (14), $\dfrac{\sinh j\theta}{\cosh j\theta} = \dfrac{j\sin\theta}{\cos\theta} = j\tan\theta$

Hence **$\tanh j\theta = j \tan\theta$** (16)

12 Differential calculus

1   Calculus is a branch of mathematics involving or leading to calculations dealing with continuously varying functions. Calculus is a subject which falls into two parts: (i) **differential calculus** (or **differentiation**) and (ii) **integral calculus** (or **integration**).

2   In an equation such as $y = 3x^2 + 2x - 5$, y is said to be a function of x and may be written as $y = f(x)$. An equation written in the form $f(x) = 3x^2 + 2x - 5$ is termed **functional notation**. The value of $f(x)$ when $x = 0$ is denoted by $f(0)$, and the value of $f(x)$ when $x = 2$ is denoted by $f(2)$, and so on.

Thus when $f(x) = 3x^2 + 2x - 5$, then $f(0) = 3(0)^2 + 2(0) - 5 = -5$,

$$\text{and } f(2) = 3(2)^2 + 2(2) - 5 = 11,$$
$$\text{and so on.}$$

3   If a tangent is drawn at a point P on a curve, then the gradient of this tangent is said to be the **gradient of the curve** at P. In *Figure 12.1*, the gradient of the curve at P is equal to the gradient of the tangent PQ.

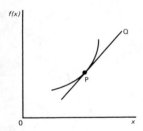

Figure 12.1

4   For the curve shown in *Figure 12.2*, let the points A and B have co-ordinates (x_1, y_1) and (x_2, y_2) respectively. In functional notation $y_1 = f(x_1)$ and $y_2 = f(x_2)$ as shown.

The gradient of the chord $AB = \dfrac{BC}{AC} = \dfrac{BD - CD}{ED} = \dfrac{f(x_2) - f(x_1)}{(x_2 - x_1)}$

162

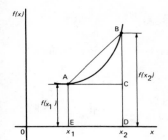

Figure 12.2

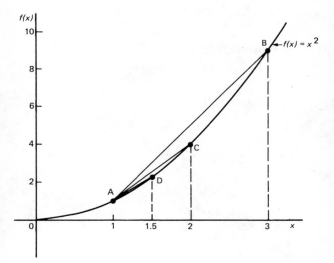

Figure 12.3

5   For the curve $f(x) = x^2$ shown in *Figure 12.3*

      (i) the gradient of chord $AB = \dfrac{f(3) - f(1)}{3 - 1} = \dfrac{9 - 1}{2} = 4,$

      (ii) the gradient of chord $AC = \dfrac{f(2) - f(1)}{2 - 1} = \dfrac{4 - 1}{1} = 3,$

(iii) the gradient of chord $AD = \dfrac{f(1.5) - f(1)}{1.5 - 1} = \dfrac{2.25 - 1}{0.5} = 2.5$,

(iv) if E is the point on the curve $(1.1, f(1.1))$ then the gradient of chord $AE = \dfrac{f(1.1) - f(1)}{1.1 - 1} = \dfrac{1.21 - 1}{0.1} = 2.1$,

(v) if F is the point on the curve $(1.01, f(1.01))$ then the gradient of chord $AF = \dfrac{f(1.01) - f(1)}{1.01 - 1} = \dfrac{1.0201 - 1}{0.01} = 2.01$.

Thus as point B moves closer and closer to point A the gradient of the chord approaches nearer and nearer to the value 2. This is called the **limiting value** of the gradient of the chord AB and when B coincides with A the chord becomes the tangent to the curve.

Differentiation from first principles

6
(i) In *Figure 12.4*, A and B are two points very close together on a curve, δx (delta x) and δy (delta y) representing small increments in the x and y directions respectively.

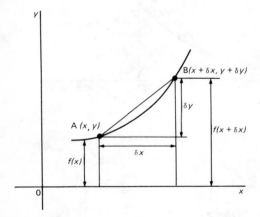

Figure 12.4

Gradient of chord $AB = \dfrac{\delta y}{\delta x}$

However $\delta y = f(x + \delta x) - f(x)$

Hence $\dfrac{\delta y}{\delta x} = \dfrac{f(x + \delta x) - f(x)}{\delta x}$

As δx approaches zero, $\dfrac{\delta y}{\delta x}$ approaches a limiting value and the gradient of the chord approaches the gradient of the tangent at A.

(ii) When determining the gradient of a tangent to a curve there are two notations used. The gradient of the curve at A in *Figure 12.4* can either by written as

$$\lim_{\delta x \to 0} \frac{\delta y}{\delta x} \text{ or } \lim_{\delta x \to 0} \left\{ \frac{f(x + \delta x) - f(x)}{\delta x} \right\}$$

In Leibniz notation, $\dfrac{dy}{dx} = \lim_{\delta x \to 0} \dfrac{dy}{dx}$

In functional notation,

$$f'(x) = \lim_{\delta_x \to 0} \left\{ \frac{f(x + \delta x) - f(x)}{\delta x} \right\}$$

(iii) $\dfrac{dy}{dx}$ is the same as $f'(x)$ and is called the **differential coefficient** or the **derivative**. The process of finding the differential coefficient is called **differentiation**. Summarising, the differential coefficient,

$$\frac{dy}{dx} = f'(x) = \lim_{\delta x \to 0} \frac{dy}{dx} = \lim_{\delta x \to 0} \left\{ \frac{f(x + \delta x) - f(x)}{\delta x} \right\}$$

For example, to differentiate $y = 3x^2$ from first principles:

$y = f(x) = 3x^2$

$f(x + \delta x) = 3(x + \delta x)^2 = 3(x^2 + 2x\delta x + \delta x^2) = 3x^2 + 6x\delta x + 3\delta x^2$

$$f'(x) = \lim_{\delta x \to 0} \left\{ \frac{f(x + \delta x) - f(x)}{\delta x} \right\} = \lim_{\delta x \to 0} \left\{ \frac{(3x^2 + 6x\delta x + 3\delta x^2) - (3x^2)}{\delta x} \right\}$$

$$= \lim_{\delta x \to 0} \left\{ \frac{6x\delta x + 3\delta x^2}{\delta x} \right\} = \lim_{\delta x \to 0} \{6x + 3\delta x\}$$

i.e., $f'(x) = 6x = \dfrac{dy}{dx}$

7 When differentiating, results can be expressed in a number of ways. For example:

(i) if $y = 3x^2$ then $\dfrac{dy}{dx} = 6x$,

(ii) if $f(x) = 3x^2$ then $f'(x) = 6x$

(iii) the differential coefficient of $3x^2$ is $6x$,

(iv) the derivative of $3x^2$ is $6x$, and

(v) $\dfrac{d}{dx}(3x^2) = 6x$.

Standard derivatives

8

y or $f(x)$	$\dfrac{dy}{dx}$ or $f'(x)$
ax^n	anx^{n-1}
$\sin ax$	$a \cos ax$
$\cos ax$	$-a \sin ax$
$\tan ax$	$a \sec^2 ax$
$\sec ax$	$a \sec ax \tan ax$
$\operatorname{cosec} ax$	$-a \operatorname{cosec} ax \cot ax$
$\cot ax$	$-a \operatorname{cosec}^2 ax$
e^{ax}	ae^{ax}
$\ln ax$	$\dfrac{1}{x}$

Thus, if $y = 4x^7$ then $\dfrac{dy}{dx} = 28x^6$,

$$y = 5\sqrt{x} = 5x^{1/2}, \ \frac{dy}{dx} = \frac{5}{2}x^{-1/2} = \frac{5}{2x^{1/2}} = \frac{5}{2\sqrt{x}},$$

$$y = \frac{6}{\sqrt[3]{x^4}} = \frac{6}{x^{4/3}} = 6x^{-4/3}, \ \frac{dy}{dx} = (6)\left(-\frac{4}{3}\right)x^{(-4/3)-1} = -8x^{-7/3}$$

$$= \frac{-8}{x^{7/3}} = -\frac{8}{\sqrt[3]{x^7}},$$

$$y = 3 \sin \omega t, \quad \frac{dy}{dt} = 3\omega \cos \omega t,$$

$$y = \frac{2}{e^{5x}} = 2e^{-5x}, \quad \frac{dy}{dx} = -10e^{-5x} = \frac{-10}{e^{5x}}$$

$$y = 4 \ln 3x, \quad \frac{dy}{dx} = \frac{4}{x}$$

9   The **differential coefficient of a sum or difference** is the sum or difference of the differential coefficients of the separate terms. Thus, if $f(x) = p(x) + q(x) - r(x)$, (where f, p, q and r are functions), then $f'(x) = p'(x) + q'(x) - r'(x)$.

Differentiation of a product

10   When $y = uv$, and u and v are both functions of x, then

$$\frac{dy}{dx} = u\frac{dv}{dx} + v\frac{du}{dx}$$

This is known as the **product rule**.
For example, if $y = 3x^2 \sin 2x$ then

$$\frac{dy}{dx} = (3x^2)(2 \cos 2x) + (\sin 2x)(6x)$$

$$= 6x^2 \cos 2x + 6x \sin 2x$$

Differentiation of a quotient

11   When $y = \dfrac{u}{v}$, and u and v are both functions of x then

$$\frac{dy}{dx} = \frac{v\dfrac{du}{dx} - u\dfrac{dv}{dx}}{v^2}$$

This is known as the **quotient rule**.
For example, if $y = \dfrac{2 \cos 3x}{x^3}$

Then, $\dfrac{dy}{dx} = \dfrac{(x^3)(-6 \sin 3x) - (2 \cos 3x)(3x^2)}{(x^3)^2}$

$$= \frac{-6x^3 \sin 3x - 6x^2 \cos 3x}{x^6} = \frac{-6x^2}{x^6}(x \sin 3x + \cos 3x)$$

$$= \frac{-6}{x^4}(x \sin 3x + \cos 3x)$$

167

Function of a function

12   It is often easier to make a substitution before differentiating.
If u is a function of x then

$$\frac{dy}{dx} = \frac{dy}{du} \times \frac{du}{dx}$$

This is known as the **'function of a function'** rule (or
sometimes, the **chain rule**).

For example, if $y = (3x-1)^9$ then, by making the substitution
$u = (3x-1)$, $y = u^9$, which is of the 'standard' form shown in para 8.
Hence

$$\frac{dy}{du} = 9u^8 \text{ and } \frac{du}{dx} = 3.$$

Then

$$\frac{dy}{dx} = \frac{dy}{du} \times \frac{du}{dx} = (9u^8)(3) = 27u^8 = 27(3x-1)^8.$$

Since y is a function of u, and u is a function of x, then y is a
function of a function of x.

Successive differentiation

13   When a function $y = f(x)$ is differentiated with respect to x the
differential coefficient is written as $\frac{dy}{dx}$ or $f'(x)$. If the expression is
differentiated again, the second differential coefficient is obtained
and is written as $\frac{d^2y}{dx^2}$ (pronounced dee two y by dee x squared) or
$f''(x)$ (pronounced f double-dash x). By successive differentiation
further higher derivatives such as $\frac{d^3y}{dx^3}$ and $\frac{d^4y}{dx^4}$ may be obtained.
Thus if

$$y = 5x^4, \frac{dy}{dx} = 20x^3, \frac{d^2y}{dx^2} = 60x^2, \frac{d^3y}{dx^3} = 120x, \frac{d^4y}{dx^4} = 120 \text{ and } \frac{d^5y}{dx^5} = 0$$

Implicit differentiation

14   When an equation can be written in the form $y = f(x)$ it is
said to be an **explicit function** of x. Examples of explicit functions

include

$$y = 2x^3 - 3x + 4, \quad y = 2x \ln x \quad \text{and} \quad y = \frac{3e^x}{\cos x}.$$

In these examples y may be differentiated with respect to x by using standard derivatives, the product rule and the quotient rule of differentiation respectively.

15   Sometimes with equations involving, say, y and x, it is impossible to make y the subject of the formula. The equation is then called an **implicit function** and examples of such functions include $y^3 + 2x^2 = y^2 - x$ and $\sin y = x^2 + 2xy$.

16   It is possible to **differentiate an implicit function** by using the function of a function rule, which may be stated as

$$\frac{du}{dx} = \frac{du}{dy} \times \frac{dy}{dx}.$$

Thus, to differentiate y^3 with respect to x, the substitution $u = y^3$ is made, from which,

$$\frac{du}{dy} = 3y^2.$$

Hence,

$$\frac{d}{dx}(y^3) = (3y^2) \times \left(\frac{dy}{dx}\right),$$

by the function of a function rule.

17   A simple rule for differentiating an implicit function is summarised as:

$$\boldsymbol{\frac{d}{dx}[f(y)] = \frac{d}{dy}[f(y)] \times \frac{dy}{dx}} \tag{1}$$

18   The product and quotient rules of differentiation must be applied when differentiating functions containing products and quotients of two variables. For example

$$\frac{d}{dx}(x^2 y) = (x^2)\frac{d}{dx}(y) + (y)\frac{d}{dx}(x^2), \qquad \text{by the product rule}$$

$$= (x^2)\left(1\frac{dy}{dx}\right) + y(2x), \qquad \text{by using equation (1)}$$

$$= \boldsymbol{x^2\frac{dy}{dx} + 2xy.}$$

19   An implicit function such as $3x^2 + y^2 - 5x + y = 2$, may be differentiated term by term with respect to x. This gives:

$$\frac{d}{dx}(3x^2) + \frac{d}{dx}(y^2) - \frac{d}{dx}(5x) + \frac{d}{dx}(y) = \frac{d}{dx}(2)$$

i.e.

$$6x + 2y\frac{dy}{dx} - 5 + 1\frac{dy}{dx} = 0,$$

using equation (1) and standard derivatives. An expression for the derivative $\frac{dy}{dx}$ in terms of x and y may be obtained by rearranging this latter equation. Thus:

$$(2y + 1)\frac{dy}{dx} = 5 - 6x$$

from which,

$$\frac{dy}{dx} = \frac{5 - 6x}{2y + 1}$$

Logarithmic differentiation

20   With certain functions containing more complicated products and quotients, differentiation is often made easier if the logarithm of the function is taken before differentiating. This technique, called **'logarithmic differentiation'** is achieved with a knowledge of

(i) the laws of logarithms,
(ii) the differential coefficients of logarithmic functions, and
(iii) the differentiation of implicit functions.

21   **Three laws of logarithms** may be expressed as:

(i) $\log(A \times B) = \log A + \log B$
(ii) $\log\left(\dfrac{A}{B}\right) = \log A - \log B$
(iii) $\log A^n = n \log A$.

22   In calculus, Naperian logarithms (i.e. logarithms to a base of 'e') are invariably used. Thus for two functions $f(x)$ and $g(x)$ the laws of logarithms may be expressed as:

(i) $\ln[f(x).g(x)] = \ln f(x) + \ln g(x)$

170

(ii) $\ln\left[\dfrac{f(x)}{g(x)}\right] = \ln f(x) - \ln g(x)$

(iii) $\ln[f(x)]^n = n \ln f(x)$

23 Taking Naperian logarithms of both sides of the equation

$$y = \frac{f(x) \cdot g(x)}{h(x)}$$

gives:

$$\ln y = \ln\left\{\frac{f(x) \cdot g(x)}{h(x)}\right\}$$

which may be simplified using the laws of logarithms given in para 22, giving:

$$\ln y = \ln f(x) + \ln g(x) - \ln h(x).$$

This latter form of the equation is often easier to differentiate.

24 (i) The differential coefficient of the logarithmic function $\ln x$ is given by:

$$\frac{d}{dx}(\ln x) = \frac{1}{x}$$

(ii) More generally, it may be shown that

$$\frac{d}{dx}[\ln f(x)] = \frac{f'(x)}{f(x)} \tag{2}$$

For example, if $y = \ln(3x^2 + 2x - 1)$ then

$$\frac{dy}{dx} = \frac{6x + 2}{3x^2 + 2x - 1}$$

Similarly, if $y = \ln(\sin 3x)$ then

$$\frac{dy}{dx} = \frac{3 \cos 3x}{\sin 3x} = 3 \cot 3x$$

(iii) By using the function of a function rule:

$$\frac{d}{dx}(\ln y) = \left(\frac{1}{y}\right)\frac{dy}{dx} \tag{3}$$

25 Differentiation of an expression such as

$$y = \frac{(1 + x)^2 \sqrt{(x - 1)}}{x \sqrt{(x + 2)}}$$

171

may be achieved by using the product and quotient rules of differentiation; however the working would be rather complicated.

With logarithmic differentiation the following procedure is adopted:

(i) Take Naperian logarithms of both sides of the equation. Thus

$$\ln y = \ln\left\{\frac{(1+x)^2\sqrt{(x-1)}}{x\sqrt{(x+2)}}\right\} = \ln\left\{\frac{(1+x)^2(x-1)^{1/2}}{x(x+2)^{1/2}}\right\}$$

(ii) Apply the laws of logarithms. Thus

$$\ln y = \ln(1+x)^2 + \ln(x-1)^{1/2} - \ln x - \ln(x+2)^{1/2},$$

by laws (i) and (ii) of para 22. i.e.,

$$\ln y = 2\ln(1+x) + \frac{1}{2}\ln(x-1) - \ln x - \frac{1}{2}\ln(x+2),$$

by law (iii) of para 22.

(iii) Differentiate each term in turn with respect to x using equations (2) and (3). Thus

$$\frac{1}{y}\frac{dy}{dx} = \frac{2}{(1+x)} + \frac{\frac{1}{2}}{(x-1)} - \frac{1}{x} - \frac{\frac{1}{2}}{(x+2)}$$

(iv) Rearrange the equation to make $\dfrac{dy}{dx}$ the subject. Thus

$$\frac{dy}{dx} = y\left\{\frac{2}{(1+x)} + \frac{1}{2(x-1)} - \frac{1}{x} - \frac{1}{2(x+2)}\right\}$$

(v) Substitute for y in terms of x. Thus

$$\frac{dy}{dx} = \frac{(1+x)^2\sqrt{(x-1)}}{x\sqrt{(x+2)}}\left\{\frac{2}{(1+x)} + \frac{1}{2(x-1)} - \frac{1}{x} - \frac{1}{2(x+2)}\right\}$$

26 Whenever an expression to be differentiated contains a term raised to a power which is itself a function of the variable, then logarithmic differentiation must be used.

For example, to find $\dfrac{dy}{dx}$ given $y = x^x$.

Taking Naperian logarithms of both sides of $y = x^x$ gives:
$$\ln y = \ln x^x = x \ln x \text{ by law (iii) of para 22.}$$

Differentiating both sides with respect to x gives:

$$\left(\frac{1}{y}\right)\frac{dy}{dx} = (x)\left(\frac{1}{x}\right) + (\ln x)(1),$$

using the product rule, i.e.

$$\frac{1}{y}\frac{dy}{dx} = 1 + \ln x, \text{ from which, } \frac{dy}{dx} = y(1 + \ln x)$$

i.e., $\dfrac{dy}{dx} = x^x(1 + \ln x)$

Partial differentiation

27   In engineering, it sometimes happens that the variation of one quantity depends on changes taking place in two, or more other quantities. For example, the volume V of a cylinder is given by $V = \pi r^2 h$. The formula will change if either radius r or height h is changed. The formula for volume may be stated mathematically as $V = f(r, h)$ which means 'V is some function of r and h'. Some other practical examples include:

(i) time of oscillation, $t = 2\pi\sqrt{\left(\dfrac{l}{g}\right)}$, i.e. $t = f(l, g)$

(ii) torque $T = I\alpha$ i.e. $T = f(I, \alpha)$

(iii) pressure of an ideal gas $p = \dfrac{mRT}{V}$, i.e. $p = f(T, V)$

(iv) resonant frequency $f_0 = \dfrac{1}{2\pi\sqrt{(LC)}}$, i.e. $f_0 = f(L, C)$,

and so on.

28   (i) When differentiating a function having two variables, one variable is kept constant and the differential coefficient of the other variable is found with respect to that variable. The differential coefficient obtained is called a **partial derivative** of the function.

(ii) A 'curly dee', ∂, is used to denote a differential coefficient in an expression containing more than one variable.

Hence if $V = \pi r^2 h$ then '$\dfrac{\partial V}{\partial r}$' means the 'partial derivative of V with respect to r, with h remaining constant'.

Thus $\dfrac{\partial V}{\partial r} = (\pi h)\dfrac{d}{dr}(r^2) = (\pi h)(2r) = 2\pi rh$.

Similarly, '$\dfrac{\partial V}{\partial h}$' means 'the partial derivative of V with respect to h, with r remaining constant'.

Thus $\dfrac{\partial V}{\partial h} = (\pi r^2)\dfrac{d}{dh}(h) = \pi r^2(1) = \pi r^2$.

(iii) $\dfrac{\partial V}{\partial r}$ and $\dfrac{\partial V}{\partial h}$ are examples of first order partial

derivatives, since $n = 1$ when written in the form $\dfrac{\partial^n V}{\partial r^n}$.

Thus, for example, if $Z = 5x^4 + 2x^3y^2 - 3y$ then

$$\dfrac{\partial Z}{\partial x} = \dfrac{d}{dx}(5x^4) + (2y^2)\dfrac{d}{dx}(x^3) - (3y)\dfrac{dy}{dx}(1)$$

$$= 20x^3 + (2y^2)(3x^2) - (3y)(0)$$

$$= \mathbf{20x^3 + 6x^2y^2}$$

and

$$\dfrac{\partial Z}{\partial y} = (5x^4)\dfrac{d}{dy}(1) + (2x^3)\dfrac{d}{dy}(y^2) - 3\dfrac{d}{dy}(y)$$

$$= (5x^4)(0) + (2x^3)(2y) - 3$$

$$= \mathbf{4x^3y - 3}.$$

29   As with ordinary differentiation, where a differential
coefficient may be differentiated again, a partial derivative may be
differentiated partially again to give higher order partial
derivatives.

(i) Differentiating $\dfrac{\partial V}{\partial r}$ of para 28 with respect to r,

keeping h constant, gives $\dfrac{\partial}{\partial r}\left(\dfrac{\partial V}{\partial r}\right)$, which is written as

$\dfrac{\partial^2 V}{\partial r^2}$.

Thus $\dfrac{\partial^2 V}{\partial r^2} = \dfrac{\partial}{\partial r}(2\pi rh) = 2\pi h$.

(ii) Differentiating $\dfrac{\partial V}{\partial h}$ with respect to h, keeping r

constant, gives

$$\dfrac{\partial}{\partial h}\left(\dfrac{\partial V}{\partial h}\right),$$

which is written as

$$\dfrac{\partial^2 V}{\partial h^2}$$

Thus

$$\frac{\partial^2 V}{\partial h^2} = \frac{\partial}{\partial h}(\pi r^2) = 0$$

(iii) Differentiating $\frac{\partial V}{\partial h}$ with respect to r, keeping h constant, gives

$$\frac{\partial}{\partial r}\left(\frac{\partial V}{\partial h}\right),$$

which is written as

$$\frac{\partial^2 V}{\partial r \partial h}$$

Thus

$$\frac{\partial^2 V}{\partial r \partial h} = \frac{\partial}{\partial r}\left(\frac{\partial V}{\partial h}\right) = \frac{\partial}{\partial r}(\pi r^2) = 2\pi r$$

(iv) Differentiating $\frac{\partial V}{\partial r}$ with respect to h, keeping r constant, gives

$$\frac{\partial}{\partial h}\left(\frac{\partial V}{\partial r}\right),$$

which is written as

$$\frac{\partial^2 V}{\partial h \partial r}.$$

Thus

$$\frac{\partial^2 V}{\partial h \partial r} = \frac{\partial}{\partial h}\left(\frac{\partial V}{\partial r}\right) = \frac{\partial}{\partial h}(2\pi r h) = 2\pi r.$$

(v) $\frac{\partial^2 V}{\partial r^2}$, $\frac{\partial^2 V}{\partial h^2}$, $\frac{\partial^2 V}{\partial r \partial h}$ and $\frac{\partial^2 V}{\partial h \partial r}$

are examples of **second order partial derivatives**.

(vi) It is seen from (iii) and (iv) that

$$\frac{\partial^2 V}{\partial r \partial h} = \frac{\partial^2 V}{\partial h \partial r}$$

and such a result is always true for continuous functions

(i.e. a graph of the function has no sudden jumps or breaks).

Thus, for example, given $Z = 4x^2y^3 - 2x^3 + 7y^2$,

(a) $\dfrac{\partial Z}{\partial x} = 8xy^3 - 6x^2$

$\dfrac{\partial^2 Z}{\partial x^2} = \dfrac{\partial}{\partial x}\left(\dfrac{\partial Z}{\partial x}\right) = \dfrac{\partial}{\partial x}(8xy^3 - 6x^2) = 8y^3 - 12x.$

(b) $\dfrac{\partial Z}{\partial y} = 12x^2y^2 + 14y$

$\dfrac{\partial^2 Z}{\partial y^2} = \dfrac{\partial}{\partial y}\left(\dfrac{\partial Z}{\partial y}\right) = \dfrac{\partial}{\partial y}(12x^2y^2 + 14y) = 24x^2y + 14.$

(c) $\dfrac{\partial^2 Z}{\partial x \partial y} = \dfrac{\partial}{\partial x}\left(\dfrac{\partial Z}{\partial y}\right) = \dfrac{\partial}{\partial x}(12x^2y^2 + 14y) = \mathbf{24xy^2}$

(d) $\dfrac{\partial^2 Z}{\partial y \partial x} = \dfrac{\partial}{\partial y}\left(\dfrac{\partial Z}{\partial x}\right) = \dfrac{\partial}{\partial y}(8xy^3 - 6x^2) = \mathbf{24xy^2}$

$\left(\text{it is noted that } \dfrac{\partial^2 Z}{\partial x \partial y} = \dfrac{\partial^2 Z}{\partial y \partial x}\right)$

30   (i) First order partial derivatives are used when finding the total differential, rates of change and errors for functions of two or more variables.

(ii) Second order partial derivatives are used in the solution of partial differential equations, in waveguide theory, and in such areas of thermodynamic covering entropy and the continuity theorem.

Total differential

31   In paras 27–29, partial differentiation is introduced for the case where only one variable changes at a time, the other variables being kept constant. In practice, variables may all be changing at the same time. If $Z = f(u, v, w, \ldots)$, i.e., the variables in the equation are $u, v, w, \ldots$, then the **total differential**, dZ, is given by the sum of the separate partial differentials of Z.

i.e. $\boxed{dZ = \dfrac{\partial Z}{\partial u}\,du + \dfrac{\partial Z}{\partial v}\,dv + \dfrac{\partial Z}{\partial w}\,dw + \ldots}$ (4)

The total differential is used as a basis for solving partial differential equations.

Thus, for example, if $Z = f(u, v, w)$ and $Z = 3u^2 - 2v + 4w^3v^2$

176

Then, the total differential $dZ = \dfrac{\partial Z}{\partial u} du + \dfrac{\partial Z}{\partial v} dv + \dfrac{\partial Z}{\partial w} dw$

$\dfrac{\partial Z}{\partial u} = 6u$ (i.e. y and w are kept constant)

$\dfrac{\partial Z}{\partial v} = -2 + 8w^3 v$ (i.e., u and w are kept constant)

$\dfrac{\partial Z}{\partial w} = 12v^2 w^2$ (i.e., u and v are kept constant)

Hence $dZ = 6u\ du + (8vw^3 - 2)dv + (12v^2 w^2)dw$.

Rates of change

32   Sometimes it is necessary to solve problems in which different quantities have different rates of change. From equation (4), the rate of change of Z, $\dfrac{dZ}{dt}$ is given by:

$$\boxed{\dfrac{dZ}{dt} = \dfrac{\partial Z}{\partial u}\dfrac{du}{dt} + \dfrac{\partial Z}{\partial v}\dfrac{dv}{dt} + \dfrac{\partial Z}{\partial w}\dfrac{dw}{dt} + \ldots} \tag{5}$$

Thus, for example, if the height of a right circular cone is increasing at 3 mm/s and its radius is decreasing at 2 mm/s then the rate of change of volume may be determined when the height is, say, 32 mm and the radius is 15 mm.

Volume of a right circular cone,

$$V = \dfrac{1}{3}\pi r^2 h$$

Using equation (5), the rate of change of volume,

$$\dfrac{dV}{dt} = \dfrac{\partial V}{\partial r}\dfrac{dr}{dt} + \dfrac{\partial V}{\partial h}\dfrac{dh}{dt}$$

$$\dfrac{\partial V}{\partial r} = \dfrac{2}{3}\pi r h \text{ and } \dfrac{\partial V}{\partial h} = \dfrac{1}{3}\pi r^2.$$

Since the height is increasing at 3 mm/s, i.e. 0.3 cm/s, then $\dfrac{dh}{dt} = +0.3$ and since the radius is decreasing at 2 mm/s, i.e.

0.2 cm/s, then $\dfrac{dr}{dt} = -0.2.$

Hence $\dfrac{dV}{dt} = \left(\dfrac{2}{3}\pi rh\right)(-0.2) + \left(\dfrac{1}{3}\pi r^2\right)(+0.3) = \dfrac{-0.4}{3}\pi rh + 0.1\pi r^2$

However, $h = 3.2$ cm and $r = 1.5$ cm.

Hence $\dfrac{dV}{dt} = \dfrac{-0.4}{3}\pi(1.5)(3.2) + (0.1)\pi(1.5)^2$

$\qquad\quad = -2.011 + 0.707 = -1.304$ cm^3/s.

Thus the rate of change of volume is 1.30 cm^3/s decreasing.

Small changes

33   It is often useful to find an approximate value for the change (or error) of a quantity caused by small changes (or errors) in the variables associated with the quantity. If $\mathcal{Z} = f(u, v, w, \ldots)$ and δu, δv, δw, $\ldots$ denote small changes in $u, v, w, \ldots$ respectively, then the corresponding approximate changes $\delta\mathcal{Z}$ in $\mathcal{Z}$ is obtained from equation (4) by replacing the differentials by the small changes. Thus

$$\boxed{\delta\mathcal{Z} \approx \dfrac{\partial\mathcal{Z}}{\partial u}\delta u + \dfrac{\partial\mathcal{Z}}{\partial v}\delta v + \dfrac{\partial\mathcal{Z}}{\partial w}\delta w + \ldots} \qquad (6)$$

Thus, for example, if the modulus of rigidity $G = \dfrac{R^4\theta}{L}$, where R is the radius, θ the angle of twist and L the length, then the approximate percentage error in G may be determined when, say, R is increased by 2%, θ is reduced by 5% and L is increased by 4%. Using equation (6), the approximate error in G, δG is given by:

$$\delta G \approx \dfrac{\partial G}{\partial R}\delta R + \dfrac{\partial G}{\partial \theta}\delta\theta + \dfrac{\partial G}{\partial L}\delta L$$

Since $G = \dfrac{R^4\theta}{L}$, $\dfrac{\partial G}{\partial R} = \dfrac{4R^3\theta}{L}$; $\dfrac{\partial G}{\partial \theta} = \dfrac{R^4}{L}$; $\dfrac{\partial G}{\partial L} = \dfrac{-R^4\theta}{L^2}$

Since R is increased by 2%, $\delta R = \dfrac{2}{100}R = 0.02R$

Similarly, $\delta\theta = -0.05\theta$ and $\delta L = 0.04L$.

Hence $\delta G \approx \left(\dfrac{4R^3\theta}{L}\right)(0.02R) + \left(\dfrac{R^4}{L}\right)(-0.05\theta) + \left(\dfrac{R^4\theta}{L^2}\right)(0.04L)$

$\qquad \approx \dfrac{R^4\theta}{L}[0.08 - 0.05 - 0.04] \approx -0.01\dfrac{R^4\theta}{L}$ i.e., $\delta G = -\dfrac{1}{100}G$

Hence the approximate percentage error in G is a 1% decrease.

Differential coefficients of inverse trigonometric functions

34

y or $f(x)$	$\dfrac{dy}{dx}$ or $f'(x)$
(i) $\arcsin \dfrac{x}{a}$	$\dfrac{1}{\surd(a^2 - x^2)}$
$\arcsin f(x)$	$\dfrac{f'(x)}{\surd\{1 - [f(x)]^2\}}$
(ii) $\arccos \dfrac{x}{a}$	$\dfrac{-1}{\surd(a^2 - x^2)}$
$\arccos f(x)$	$\dfrac{-f'(x)}{\surd\{1 - [f(x)]]^2\}}$
(iii) $\arctan \dfrac{x}{a}$	$\dfrac{a}{a^2 + x^2}$
$\arctan f(x)$	$\dfrac{f'(x)}{\{1 + [f(x)]^2\}}$
(iv) $\text{arcsec} \dfrac{x}{a}$	$\dfrac{a}{x\surd(x^2 - a^2)}$
$\text{arcsec} f(x)$	$\dfrac{f'(x)}{f(x)\surd\{[f(x)]^2 - 1\}}$
(v) $\text{arccosec} \dfrac{x}{a}$	$\dfrac{-a}{x\surd(x^2 - a^2}$
$\text{arccosec} f(x)$	$\dfrac{-f'(x)}{f(x)\surd\{[f(x)]^2 - 1\}}$
(vi) $\text{arccot} \dfrac{x}{a}$	$\dfrac{-a}{a^2 + x^2}$
$\text{arccot} f(x)$	$\dfrac{-f'(x)}{\{1 + [f(x)]^2\}}$

Thus, for example, if $y = \arcsin 5x^2$ then

$$\frac{dy}{dx} = \frac{10x}{\sqrt{(1 - 25x^4)}}$$

Similarly, if $y = \arctan \dfrac{3}{t^2}$ then

$$\frac{dy}{dt} = \frac{(-6/t^3)}{\left(1 + \dfrac{9}{t^4}\right)} = \frac{(-6/t^3)}{\left(\dfrac{t^4 + 9}{t^4}\right)} = \frac{-6t}{t^4 + 9}$$

Differential coefficients of hyperbolic functions

35

y or $f(x)$	$\dfrac{dy}{dx}$ or $f'(x)$
sinh ax	a cosh ax
cosh ax	a sinh ax
tanh ax	a sech^2 ax
sech ax	$-a$ sech ax tanh ax
cosech ax	$-a$ cosech ax coth ax
coth ax	$-a$ cosech^2 ax

Thus, for example if $y = 4 \text{ sh } 2x - \dfrac{3}{7} \text{ ch } 3x$

$$\text{then } \frac{dy}{dx} = 4(2 \text{ ch } 2x) - \frac{3}{7}(3 \text{ sh } 3x)$$

$$= \mathbf{8 \text{ ch } 2x - \frac{9}{7} \text{ sh } 3x}$$

Similarly, if $y = 5 \text{ th} \dfrac{x}{2} - 2 \text{ coth } 4x$

$$\text{then } \frac{dy}{dx} = 5\left(\frac{1}{2}\text{sech}^2\frac{x}{2}\right) - 2(-4 \text{ cosech}^2 \, 4x)$$

$$= \mathbf{\frac{5}{2} \text{ sech}^2\frac{x}{2} + 8 \text{ cosech}^2 \, 4x.}$$

Differential coefficients of inverse hyperbolic functions

y or $f(x)$	$\dfrac{dy}{dx}$ or $f'(x)$
(i) $\quad \text{arsinh} \dfrac{x}{a}$	$\dfrac{1}{\sqrt{(x^2 + a^2)}}$
$\quad \text{arsinh}\, f(x)$	$\dfrac{f'(x)}{\sqrt{\{[f(x)]^2 + 1\}}}$
(ii) $\quad \text{arcosh} \dfrac{x}{a}$	$\dfrac{1}{\sqrt{(x^2 - a^2)}}$
$\quad \text{arcosh}\, f(x)$	$\dfrac{f'(x)}{\sqrt{\{[f(x)]^2 - 1\}}}$
(iii) $\quad \text{artanh} \dfrac{x}{a}$	$\dfrac{a}{a^2 - x^2}$
$\quad \text{artanh}\, f(x)$	$\dfrac{f'(x)}{\{1 - [f(x)]^2\}}$
(iv) $\quad \text{arsech} \dfrac{x}{a}$	$\dfrac{-a}{x\sqrt{(a^2 - x^2)}}$
$\quad \text{arsech}\, f(x)$	$\dfrac{-f'(x)}{f(x)\sqrt{\{1 - [f(x)]^2\}}}$
(v) $\quad \text{arcosech} \dfrac{x}{a}$	$\dfrac{-a}{x\sqrt{(x^2 + a^2)}}$
$\quad \text{arcosech}\, f(x)$	$\dfrac{-f'(x)}{f(x)\sqrt{\{[f(x)]^2 + 1\}}}$
(vi) $\quad \text{arcoth} \dfrac{x}{a}$	$\dfrac{a}{a^2 - x^2}$
$\quad \text{arcoth}\, f(x)$	$\dfrac{f'(x)}{\{1 - [f(x)]^2\}}$

Thus, for example, if $y = \operatorname{arcosh}(5x^2 - 1)$ then

$$\frac{dy}{dx} = \frac{10x}{\sqrt{[(5x^2-1)^2-1]}} = \frac{10x}{\sqrt{[25x^4-10x^2]}} = \frac{10x}{\sqrt{[5x^2(5x^2-2)]}}$$

$$= \frac{10}{\sqrt{[5(5x^2-2)]}}$$

Similarly, if $y = \operatorname{arcoth}(\sin x)$ then

$$\frac{dy}{dx} = \frac{\cos x}{1-\sin^2 x} = \frac{\cos x}{\cos^2 x} = \frac{1}{\cos x} = \mathbf{sec\ } x$$

Applications of differentiation

Velocity and acceleration

37   If a body moves a distance x metres in a time t seconds then:

   (i) distance, $x = f(t)$

   (ii) velocity, $v = f'(t)$ or $\dfrac{dx}{dt}$, which is the gradient of the distance/time graph,

   (iii) acceleration, $a = \dfrac{dv}{dt} = f''(t)$ or $\dfrac{d^2x}{dt^2}$, which is the gradient of the velocity/time graph.

Thus, for example, if the distance x metres moved by a missile in a time t seconds is given by: $x = 3t^3 - 2t^2 + 4t - 1$

then velocity, $v = \dfrac{dx}{dt} = 9t^2 - 4t + 4$

and acceleration, $a - \dfrac{d^2x}{dt^2} = 18t - 4$

Thus after a time of 1.5 s, velocity $v = 9(1.5)^2 - 4(1.5) + 4 = 18.25$ m/s

and acceleration $a = 18(1.5) - 4 = \mathbf{23\ m/s^2}$.

Rates of change

38   (i) If a quantity y depends on and varies with a quantity x then the rate of change of y with respect to x is $\dfrac{dy}{dx}$. Thus, for example, the rate of change of pressure p with height h is $\dfrac{dp}{dh}$.

   (ii) A rate of change with respect to time is usually just called 'the rate of change', the 'with respect to time' being assumed. Thus, for example, a rate of change of voltage, v

is $\dfrac{dv}{dt}$ and a rate of change of temperature, θ is $\dfrac{d\theta}{dt}$, and so on.

Turning points

39 (i) In *Figure 12.5* the gradient (or rate of change) of the curve changes from positive between O and P to negative between P and Q, and then positive again between Q and R. At point P, the gradient is zero and, as x increases, the

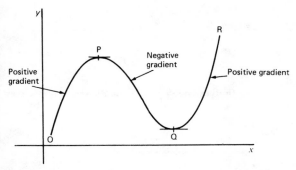

Figure 12.5

gradient of the curve changes from positive just before P to negative just after. Such a point is called a **maximum point** and appears as the 'crest of a wave'. At point Q the gradient is also zero and, as x increases, the gradient of the curve changes from negative just before Q to positive just after. Such a point is called a **minimum point** and appears as the 'bottom of a valley'. Points such as P and Q are given the general name of turning points.

(ii) It is possible to have a turning point, the gradient on either side of which is the same. Such a point is given the special name of a **point of inflexion**, and examples are shown in *Figure 12.6*.

(iii) Maximum and minimum points and points of inflexion are given the general term of **stationary points**.

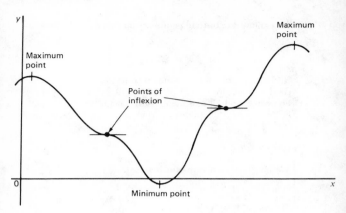

Figure 12.6

Procedure for finding and distinguishing between stationary points

40   (i) Given $y=f(x)$, determine $\dfrac{dy}{dx}$ (i.e., $f'(x)$).

(ii) Let $\dfrac{dy}{dx}=0$ and solve for the values of x.

(iii) Substitute the values of x into the original equation, $y=f(x)$, to find the corresponding y-ordinate values. This establishes the co-ordinates of the stationary points.
To determine the nature of the stationary points:
either
(iv) Find $\dfrac{d^2y}{dx^2}$ and substitute into it the values of x found

in (ii). If the result is:
    (a) positive — the point is a minimum one,
    (b) negative — the point is a maximum one,
    (c) zero — the point is a point of inflexion
         or
(v) Determine the sign of the gradient of the curve just before and just after the stationary points. If the sign change for the gradient of the curve is:
    (a) positive to negative — the point is a maximum one,
    (b) negative to positive — the point is a minimum one,
    (c) positive to positive, or negative to negative — the point is a point of inflexion.

184

For example, to find the turning points on the curve $y = x^3 - 3x + 5$:

(i) $\dfrac{dy}{dx} = 3x^2 - 3$

(ii) $3x^2 - 3 = 0$, hence $x = \pm 1$.

(iii) When $x = \pm 1$, $y = 1^3 - 3(1) + 5 = 3$

When $x = -1$, $y = (-1)^3 - 3(-1) + 5 = 7$

Thus the turning points occur at $(1, 3)$ and $(-1, 7)$

(iv) Since $\dfrac{dy}{dx} = 3x^2 - 3$, $\dfrac{d^2y}{dx^2} = 6x$

When $x = +1$, $\dfrac{d^2y}{dx^2}$ is positive.

Hence (1, 3) is a minimum point.

When $x = -1$, $\dfrac{d^2y}{dx^2}$ is negative.

Hence (−1, 7), is a maximum point.

13 Integral calculus

1   The process of integration reverses the process of differentiation. In differentiation, if $f(x) = 2x^2$ then $f'(x) = 4x$. Thus the integral of $4x$ is $2x^2$, i.e., integration is the process of moving from $f'(x)$ to $f(x)$. By similar reasoning, the integral of $2t$ is t^2.

2   Integration is also a process of summation or adding parts together and an elongated S, shows as $\int$, is used to replace the words 'the integral of'. Hence, from para 1, $\int 4x = 2x^2$ and $\int 2t = t^2$.

3   In differentiation, the differential coefficient $\dfrac{dy}{dx}$ indicates that a function of x is being differentiated with respect to x, the dx indicating that it is 'with respect to x'. In integration the variable of integration is shown by adding d (the variable) after the function to be integrated. Thus $\int 4x \, dx$ means 'the integral of $4x$ with respect to x', and $\int 2t \, dt$ means 'the integral of $2t$ with respect to t'.

4   As stated in para 1, the differential coefficient of $2x^2$ is $4x$, hence $\int 4x \, dx = 2x^2$. However, the differential coefficient of $(2x^2 + 7)$ is also $4x$. Hence $\int 4x \, dx$ is also equal to $(2x^2 + 7)$. To allow for the possible presence of a constant, whenever the process of integration is performed, a constant 'c' is added to the result. Thus $\int 4x \, dx = 2x^2 + c$ and $\int 2t \, dt = t^2 + c$. 'c' is called the **arbitrary constant of integration**.

5   The general solution of integrals of the form $\int ax^n \, dx$, where a and n are constants is given by:

$$\int ax^n \, dx = \frac{ax^{n+1}}{n+1} + c$$

This rule is true when n is fractional, zero, or a positive or negative integer, with the exception of $n = -1$. Using this rule gives:

(i) $\displaystyle\int 3x^4 \, dx = \frac{3x^{4+1}}{4+1} + c = \frac{3}{5}x^5 + c,$

(ii) $\displaystyle\int \frac{2}{x^2} \, dx = \int 2x^{-2} \, dx = \frac{2x^{-2+1}}{-2+1} + c = \frac{2x^{-1}}{-1} + c = \frac{-2}{x} + c,$

and

(iii) $\int \sqrt{x}\ dx = \int x^{1/2}\ dx = \dfrac{x^{(1/2)+1}}{(1/2)+1} + c = \dfrac{x^{3/2}}{\dfrac{3}{2}} + c = \dfrac{2}{3}\sqrt{x^3} + c$

Each of these three results may be checked by differentiation.

6   (i) The integral of a constant k is $kx + c$. For example, $\int 8dx = 8x + c$.

(ii) When a sum of several terms is integrated the result is the sum of the integrals of the separate terms. For example,

$$\int (3x + 2x^2 - 5)\,dx = \int 3x\ dx + \int 2x^2\ dx - \int 5\ dx$$
$$= \dfrac{3x^2}{2} + \dfrac{2x^3}{3} - 5x + c$$

7   Since integration is the reverse process of differentiation the **standard integrals** listed below may be deduced and readily checked by differentiation.

Standard integrals

(i)	$\int a\ x^n\ dx$	$= \dfrac{ax^{n+1}}{n+1} + c$ (except when $n = -1$)
(ii)	$\int \cos ax\ dx$	$= \dfrac{1}{a} \sin ax + c$
(iii)	$\int \sin ax\ dx$	$= -\dfrac{1}{a}\cos ax + c$
(iv)	$\int \sec^2 ax\ dx$	$= \dfrac{1}{a} \tan ax + c$
(v)	$\int \operatorname{cosec}^2 ax\ dx$	$= -\dfrac{1}{a}\cot ax + c$
(vi)	$\int \operatorname{cosec} ax \cot ax\ dx$	$= -\dfrac{1}{a}\operatorname{cosec} ax + c$
(vii)	$\int \sec ax \tan ax\ dx$	$= \dfrac{1}{a}\sec ax + c$
(viii)	$\int e^{ax}\ dx$	$= \dfrac{1}{a}e^{ax} + c$
(ix)	$\int \dfrac{1}{x}\ dx$	$= \ln x + c$

Thus, for example,

$$\int 3 \sin 2x \, dx = -\frac{3}{2} \cos 2x + c$$

$$\int \frac{5}{2} e^{3x} \, dx = \frac{5}{2(3)} e^{3x} + c = \frac{5}{6} e^{3x} + c$$

and

$$\int \frac{6}{5x} \, dx = \frac{6}{5} \int \frac{1}{x} \, dx = \frac{6}{5} \ln x + c$$

8   Integrals containing an arbitrary constant c in their results are called **indefinite integrals** since their precise value cannot be determined without further information. **Definite integrals** are those in which limits are applied. If an expression is written as $[x]_a^b$, 'b' is called the upper limit and 'a' the lower limit. The operation of applying the limits is defined as $[x]_a^b = (b - a)$. The increase in the value of the integral x^2 as x increases from 1 to 3 is written as $\int_1^3 x^2 \, dx$. Applying the limits gives:

$$\int_1^3 x^2 \, dx = \left[\frac{x^3}{3} + c\right]_1^3 = \left\{\frac{(3)^3}{3} + c\right\} - \left\{\frac{(1)^3}{3} + c\right\} = (9 + c) - \left(\frac{1}{3} + c\right)$$

$$= 8\frac{2}{3}$$

Note that the 'c' term always cancels out when limits are applied and the arbitrary constant need not be shown with definite integrals. Similarly,

$$\int_0^{\pi/6} 2 \cos 3x \, dx = \left[\frac{2}{3} \sin 3x\right]_0^{\pi/6} = \frac{2}{3}\left[\sin 3\frac{\pi}{6} - \sin 0\right] = \frac{2}{3}(1 - 0) = \frac{2}{3}$$

9   Functions which require integrating are not always in the 'standard' form shown in para 7. However, it is often possible to change a function into a form which can be integrated by using either:

        (i)   an algebraic substitution,
        (ii) a trigonometric or hyperbolic substitution,
        (iii) partial fractions or
        (iv) integration by parts.

10   With **algebraic substitutions**, the substitution usually made is to let u be equal to $f(x)$ such that $f(u)\ du$ is a standard integral. It is found that integrals of the form $k\int [f(x)]^n f'(x)\ dx$ and $k\int \dfrac{f'(x)}{[f(x)]^n}\ dx$ (where k and n are constants) can both be integrated by substituting u for $f(x)$. For example, to determine $\int (5x+2)^8\ dx$,

let $u = 5x+2$, then $\dfrac{du}{dx} = 5$ and $dx = \dfrac{du}{5}$

Hence

$$\int (5x+2)^8\ dx = \int u^8 \dfrac{du}{5} = \dfrac{1}{5}\int u^8\ du = \dfrac{1}{5}\left(\dfrac{u^9}{9}\right) + c$$

$$= \dfrac{1}{45}u^9 + c$$

Rewriting u as $(5x+2)$ gives

$$\int (5x+2)^8\ dx = \dfrac{1}{45}(5x+2)^9 + c$$

Similarly, to determine

$$\int \dfrac{1}{7y+4}\ dy$$

let $u = (7y+4)$, then $\dfrac{du}{dy} = 7$ and $dy = \dfrac{du}{7}$.

Hence

$$\int \dfrac{1}{(7y+4)}\ dy = \int \left(\dfrac{1}{u}\right)\dfrac{du}{7} = \dfrac{1}{7}\int \dfrac{1}{u}\ du = \dfrac{1}{7}\ln u + c = \dfrac{1}{7}\ln(7y+4) + c.$$

Similarly, to determine

$$\int 6e^{5t-1}\ dt,$$

let $u = (5t-1)$, then $\dfrac{du}{dt} = 5$ and $dt = \dfrac{du}{5}$

Hence

$$\int 6e^{5t-1}\ dt = \int 6e^u \dfrac{du}{5} = \dfrac{6}{5}\int e^u\ du = \dfrac{6}{5}e^u + c = \dfrac{6}{5}e^{5t-1} + c.$$

Also, to determine $\int 2\sin^3\theta \cos\theta\ d\theta$,

189

let $u = \sin\theta$, then $\dfrac{du}{d\theta} = \cos\theta$ and $d\theta = \dfrac{du}{\cos\theta}$

Hence

$$\int 2\sin^3\theta \, \cos\theta \, d\theta = \int 2(u)^3 \cos\theta \dfrac{du}{\cos\theta} = 2\int u^3 \, du,$$

by cancelling,

$$= 2\left(\dfrac{u^4}{4}\right) + c = \dfrac{1}{2}\sin^4\theta + c.$$

(Note that $\sin^3\theta \equiv (\sin\theta)^3$).

To determine $\int 3x(2x^2 - 5)^5 \, dx$,

let $u = (2x^2 - 5)$, then $\dfrac{du}{dx} = 4x$ and $dx = \dfrac{du}{4x}$

Hence

$$\int 3x(2x^2 - 5)^5 \, dx = \int 3x(u)^5 \dfrac{du}{4x} = \dfrac{3}{4}\int u^5 \, du,$$

by cancelling.

The original variable 'x', has been completely removed, and the integral is now only in terms of u and is a 'standard integral'.

Hence

$$\dfrac{3}{4}\int u^5 \, du = \dfrac{3}{4}\left(\dfrac{u^6}{6}\right) + c = \dfrac{3}{24}(2x^2 - 5)^6 + c = \dfrac{1}{8}(2x^2 - 5)^6 + c.$$

To determine

$$\int \dfrac{3x}{\sqrt{(4x^2 - 3)}} \, dx,$$

let $u = (4x^2 - 3)$, then $\dfrac{du}{dx} = 8x$ and $dx = \dfrac{du}{8x}$.

Hence

$$\int \dfrac{3x}{\sqrt{(4x^2 - 3)}} \, dx = \int \dfrac{3x}{\sqrt{u}}\left(\dfrac{du}{8x}\right) = \dfrac{3}{8}\int \dfrac{1}{\sqrt{u}} \, du = \dfrac{3}{8}\int u^{-1/2} \, du$$

$$= \left(\dfrac{3}{8}\right)\dfrac{u^{-\frac{1}{2}+1}}{-\dfrac{1}{2}+1} + c = \left(\dfrac{3}{8}\right)\dfrac{u^{1/2}}{\dfrac{1}{2}} + c = \dfrac{3}{4}\sqrt{u} + c$$

$$= \dfrac{3}{4}\sqrt{(4x^2 - 3)} + c$$

11   *Table 13.1* summarises the integrals that require the
use of a **trigonometric or hyperbolic substitution**.
Thus, for example,

$$\int 2 \cos^2 4t \; dt = 2 \int \frac{1}{2}(1 + \cos 8t) \; dt, \text{ from 1 of } Table \; 13.1$$

$$= t + \frac{\sin 8t}{8} + c$$

$$\int \sin^2 x \cos^3 x \; dx = \int \sin^2 x \cos^2 x \cos x \; dx$$

$$= \int \sin^2 x (1 - \sin^2 x) \cos x \; dx, \text{ from 5(a) of } Table \; 13.1$$

$$= \int (\sin^2 x \cos x - \sin^4 x \cos x) \; dx$$

$$= \frac{\sin^3 x}{3} - \frac{\sin^5 x}{5} + c$$

$$\int \sin^2 t \cos^4 t \; dt = \int \sin^2 t \; (\cos^2 t)^2 \; dt$$

$$= \int \left(\frac{1 - \cos 2t}{2}\right)\left(\frac{1 + \cos 2t}{2}\right)^2 \; dt,$$
from 5(b) of *Table 13.1*

$$= \frac{1}{8}\int (1 - \cos 2t)(1 + 2 \cos 2t + \cos^2 2t) \; dt$$

$$= \frac{1}{8}\int (1 + 2 \cos 2t + \cos^2 2t - \cos 2t - 2 \cos^2 2t - \cos^2 2t) \; dt$$

$$= \frac{1}{8}\int (1 + \cos 2t - \cos^2 2t - \cos^3 2t) \; dt$$

$$= \frac{1}{8}\int \left[1 + \cos 2t - \left(\frac{1 + \cos 4t}{2}\right) - \cos 2t(1 - \sin^2 2t) \right] dt$$

$$= \frac{1}{8}\int \left(\frac{1}{2} - \frac{\cos 4t}{2} + \cos 2t \sin^2 2t \right) dt$$

$$= \frac{1}{8}\left(\frac{t}{2} - \frac{\sin 4t}{8} + \frac{\sin^3 2t}{6}\right) + c$$

$$\int \frac{1}{3}\cos 5x \sin 2x \; dx = \frac{1}{3}\int \frac{1}{2}[\sin(5x + 2x) - \sin(5x - 2x)] \; dx, \text{ from 7}$$
of *Table 13.1*

$$= \frac{1}{6}\int (\sin 7x - \sin 3x) \; dx = \frac{1}{6}\left(-\frac{\cos 7x}{7} + \frac{\cos 3x}{3}\right) + c$$

Table 13.1

$f(x)$	$\int f(x)\, dx$	Method
1 $\cos^2 x$	$\dfrac{1}{2}\left(x+\dfrac{\sin 2x}{2}\right)+c$	Use $\cos 2x = 2\cos^2 x - 1$
2 $\sin^2 x$	$\dfrac{1}{2}\left(x-\dfrac{\sin 2x}{2}\right)+c$	Use $\cos 2x = 1 - 2\sin^2 x$
3 $\tan^2 x$	$\tan x - x + c$	Use $1 + \tan^2 x = \sec^2 x$
4 $\cot^2 x$	$-\cot x - x + c$	Use $\cot^2 x + 1 = \mathrm{cosec}^2 x$
5 $\cos^m x \sin^n x$	(a) If either m or n is odd (but not both), use $\cos^2 x + \sin^2 x = 1$ (b) If both m and n are even, use either $\cos 2x = 2\cos^2 x - 1$ or $\cos 2x = 1 - 2\sin^2 x$	
6 $\sin A \cos B$	Use $\dfrac{1}{2}[\sin(A+B) + \sin(A-B)]$	
7 $\cos A \sin B$	Use $\dfrac{1}{2}[\sin(A+B) - \sin(A-B)]$	
8 $\cos A \cos B$	Use $\dfrac{1}{2}[\cos(A+B) - \cos(A-B)]$	
9 $\sin A \sin B$	Use $-\dfrac{1}{2}[\cos(A+B) - \cos(A-B)]$	
10 $\dfrac{1}{\sqrt{(a^2-x^2)}}$	$\arcsin\dfrac{x}{a}+c$	Use $x = a\sin\theta$ substitution
11 $\sqrt{(a^2-x^2)}$	$\dfrac{a^2}{2}\arcsin\dfrac{x}{a}+\dfrac{x}{2}\sqrt{(a^2-x^2)}+c$	
12 $\dfrac{1}{a^2+x^2}$	$\dfrac{1}{a}\arctan\dfrac{x}{a}+c$	Use $x = a\tan\theta$ substitution
13 $\dfrac{1}{\sqrt{(x^2+a^2)}}$	$\mathrm{arsinh}\dfrac{x}{a}+c$ or $\ln\left\{\dfrac{x+\sqrt{(x^2+a^2)}}{a}\right\}+c$	Use $x = a\sinh\theta$ substitution
14 $\sqrt{(x^2+a^2)}$	$\dfrac{a^2}{2}\mathrm{arsinh}\dfrac{x}{a}+\dfrac{x}{2}\sqrt{(x^2+a^2)}+c$	
15 $\dfrac{1}{\sqrt{(x^2-a^2)}}$	$\mathrm{arcosh}\dfrac{x}{a}+c$ or $\ln\left\{\dfrac{x+\sqrt{(x^2-a^2)}}{a}\right\}+c$	Use $x = a\cosh\theta$ substitution
16 $\sqrt{(x^2-a^2)}$	$\dfrac{x}{2}\sqrt{(x^2-a^2)}-\dfrac{a^2}{2}\mathrm{arcosh}\dfrac{x}{a}+c$	

$$\int_0^9 \frac{1}{\sqrt{(9-x^2)}}\,dx = \left[\arcsin\frac{x}{3}\right]_0^3, \text{ from 10 of } Table\ 13.1$$

$$= (\arcsin 1 - \arcsin 0) = \frac{\pi}{2} \textbf{ or 1.5708}$$

$$\int_0^4 \sqrt{(16-x^2)}\,dx = \left[\frac{16}{2}\arcsin\frac{x}{4} + \frac{x}{2}\sqrt{(16-x^2)}\right]_0^4, \text{ from 11 of } Table\ 13.1$$

$$= (8\arcsin 1 + 2\sqrt{0}) - (8\arcsin 0 + 0)$$

$$= 8\arcsin 1 = 8\left(\frac{\pi}{2}\right) = \textbf{4}\pi \textbf{ or 12.57}$$

12   The process of expressing a fraction in terms of simpler fractions — called **partial fractions** — is discussed in the section on Algebra (Chapter 3) and the forms of partial fractions used are summarised on page 35. Thus, for example,

$$\int \frac{11-3x}{x^2+2x-3}\,dx = \int\left(\frac{2}{x-1} - \frac{5}{x+3}\right)dx \text{ (see page 36)}$$

$$= \textbf{2}\ln(\textbf{x}-\textbf{1}) - \textbf{5}\ \textbf{ln}(\textbf{x}+\textbf{3}) + \textbf{c}$$

$$\int \frac{2x+3}{(x-2)^2}\,dx = \int\left(\frac{2}{x-2} + \frac{7}{(x-2)^2}\right)dx \text{ (see page 36)}$$

$$= \textbf{2}\ \textbf{ln}(\textbf{x}-\textbf{2}) - \frac{\textbf{7}}{(\textbf{x}-\textbf{2})} + \textbf{c}$$

$$\int \frac{3+6x+4x^2-2x^2}{x^2(x^2+3)}\,dx = \int\left(\frac{2}{x} + \frac{1}{x^2} + \frac{3-4x}{x^2+3}\right)dx \text{ (see page 37)}$$

$$= \int\left(\frac{2}{x} + \frac{1}{x^2} + \frac{3}{x^2+3} - \frac{4x}{x^2+3}\right)dx$$

$$= \textbf{2}\ \textbf{ln}\ \textbf{x} - \frac{\textbf{1}}{\textbf{x}} + \frac{\textbf{3}}{\sqrt{\textbf{3}}}\ \textbf{arctan}\frac{\textbf{x}}{\sqrt{\textbf{3}}} - \textbf{2}\ \textbf{ln}(\textbf{x}^2+\textbf{3}) + \textbf{c}$$

Integration by parts

13   From the product rule of differentiation:

$$\frac{d}{dx}(uv) = v\frac{du}{dx} + u\frac{dv}{dx},$$

where u and v are both functions of x. Rearranging gives:

$$u\frac{dv}{dx} = \frac{d}{dx}(uv) - v\frac{du}{dx}$$

Integrating both sides with respect to x gives:

$$\int u\frac{dv}{dx}\,dx = \int\frac{d}{dx}(uv)\,dx - \int v\frac{du}{dx}\,dx$$

i.e. $\int u\dfrac{dv}{dx}\,dx = uv - \int v\dfrac{du}{dx}\,dx$

or **$\int u\,dv = uv - \int v\,du$**

This is known as the **integration by parts formula** and provides a method of integrating such products of simple functions as $\int xe^x\,dx$, $\int t\sin t\,dt$, $\int e^{\theta}\cos\theta\,d\theta$ and $\int x\ln x\,dx$.

14   Given a product of two terms to integrate the initial choice is: 'which part to make equal to u' and 'which part to make equal to v'. The choice must be such that the 'u part' becomes a constant after successive differentiation and the 'dv part' can be integrated from standard integrals.

Invariably, the following rule holds: 'If a product to be integrated contains an algebraic term (such as x, t^2 or 3θ) then this term is chosen as the u part'. The one exception to this rule is when a '$\ln x$' term is involved; in this case $\ln x$ is chosen as the 'u part'. Thus, for example, to determine

$$\int x\cos x\,dx,$$

let $u = x$, from which $\dfrac{du}{dx} = 1$, i.e., $du = dx$

and let $dv = \cos x\,dx$, from which, $v = \int\cos x\,dx = \sin x$. Expressions for u, du, v and dv are now substituted into the parts formula as shown below.

$$\int\boxed{\begin{matrix}u\\x\end{matrix}}\ \boxed{\begin{matrix}dv\\\cos x\,dx\end{matrix}} = \boxed{\begin{matrix}u\\(x)\end{matrix}}\ \boxed{\begin{matrix}v\\(\sin x)\end{matrix}} - \int\boxed{\begin{matrix}v\\(\sin x)\end{matrix}}\ \boxed{\begin{matrix}du\\(dx)\end{matrix}}$$

i.e.

$$\int x \cos x \ dx = x \sin x - (-\cos x) + c$$

$$= x \sin x + \cos x + c.$$

(This result may be checked by differentiating the right hand side, i.e.

$$\frac{d}{dx}(x \sin x + \cos x + c) = [(x)(\cos x) + (\sin x)(1)] - \sin x + 0$$

$$= x \cos x, \text{ which is the function being integrated.})$$

When determining $\int x \ln x \ dx$, the logarithmic function is chosen as the 'u part'. Thus when

$$u = \ln x, \text{ then } \frac{du}{dx} = \frac{1}{x}, \text{ i.e., } du = \frac{dx}{x}.$$

Letting $dv = x \ dx$ gives $v = \int x \ dx = \frac{x^2}{2}$

Substituting into $\int u \ dv = uv - \int v \ du$ gives:

$$\int x \ln x \ dx = (\ln x)\left(\frac{x^2}{2}\right) - \int\left(\frac{x^2}{2}\right)\frac{dx}{x}$$

$$= \frac{x^2}{2}\ln x - \frac{1}{2}\int x \ dx = \frac{x^2}{2}\ln x - \frac{1}{2}\left(\frac{x^2}{2}\right) + c$$

Hence

$$\int x \ln x \ dx = \frac{x^2}{2}\left(\ln x - \frac{1}{2}\right) + c \text{ or } \frac{x^2}{4}(2 \ln x - 1) + c.$$

Applications of integration

(a) Areas under and between curves

15   (i) Let A be the area shown shaded in *Figure 13.1* and let this area be divided into a number of strips each of width δx. One such strip is shown and let the area of this strip be δA. Then:

$$\delta A \approx y \ \delta x \qquad\qquad (1)$$

The accuracy of statement (1) increases when the width of each strip is reduced, i.e. area A is divided into a greater number of strips.

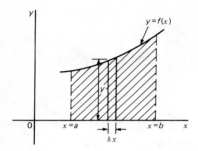

Figure 13.1

(ii) Area A is equal to the sum of all the strips from $x = a$ to $x = b$, i.e.

$$A = \frac{\text{limit}}{\delta x \to 0} \sum_{x=a}^{x=b} y \delta x \qquad (2)$$

(iii) From statement (1),

$$\frac{\delta A}{\delta x} \approx y \qquad (3)$$

In the limit, as δx approaches zero, $\dfrac{\delta A}{\delta x}$ becomes the differential coefficient $\dfrac{dA}{dx}$. Hence $\dfrac{\text{limit}}{\delta x \to 0}\left(\dfrac{\delta A}{\delta x}\right) = \dfrac{dA}{dx} = y$, from statement (3). By integration, $\int \dfrac{dA}{dx}\, dx = \int y\, dx$, i.e. $A = \int y\, dx$. The ordinates $x = a$ and $x = b$ limit the area and such ordinate values are shown as limits. Hence

$$A = \int_{a}^{b} y\, dx \qquad (4)$$

(iv) Equating statements (2) and (4) gives:

Shaded area of *Figure 13.1*, $A = \dfrac{\text{limit}}{\delta x \to 0} \sum_{x=a}^{x=b} y \delta x = \int_{n}^{b} y\, dx = \int_{a}^{b} f(x)\, dx$

196

(v) If the area between a curve $x = f(y)$, the y-axis and ordinates $y = p$ and $y = q$ is required then **area** $= \displaystyle\int_{p}^{q} x \; dy$.

Thus determining the area under a curve by integration merely involves evaluating a definite integral.

16   There are several instances in engineering and science where the area beneath a curve needs to be accurately determined. For example, the areas between limits of a:

velocity/time graph gives distance travelled,

force/distance graph gives work done,

voltage/current graph gives power, and so on.

17   Should a curve drop below the x-axis, then $y \, (= f(x))$ becomes negative and $\int f(x) \; dx$ is negative. When determining such areas by integration, a negative sign is placed before the integral. For the curve shown in *Figure 13.2*, the total shaded area is given by (area E + area F + area G).

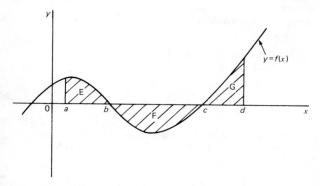

Figure 13.2

By integration,

total shaded area $= \displaystyle\int_{a}^{b} f(x) \; dx - \int_{b}^{c} f(x) \; dx + \int_{c}^{d} f(x) \; dx.$

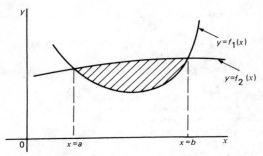

Figure 13.3

(Note that this is **not** the same as $\int\limits_{a}^{d} f(x) \ dx$.)

It is usually necessary to sketch a curve in order to check whether it crosses the x-axis.

18   The area enclosed between curves $y = f_1(x)$ and $y = f_2(x)$ (shown shaded in *Figure 13.3*) is given by:

$$\textbf{Shaded area} = \int\limits_{a}^{b} f_2(x) \ dx - \int\limits_{a}^{b} f_1(x) \ dx$$

$$= \int\limits_{a}^{b} [\boldsymbol{f_2(x) - f_1(x)}] \ \boldsymbol{dx}$$

Thus, for example, to determine the area enclosed by the curve $y = x^3 + 2x^2 - 5x - 6$ and the x-axis (i.e., the shaded area shown in *Figure 13.4*):

Shaded area $= \int\limits_{-3}^{-1} y \ dx - \int\limits_{-1}^{2} y \ dx$, the minus sign before the second

integral being necessary since the enclosed area is below the x-axis. Hence

$$\text{shaded area} = \int\limits_{-3}^{-1} (x^3 + 2x^2 - 5x - 6) \ dx - \int\limits_{1}^{2} (x^3 + 2x^2 - 5x - 6) \ dx$$

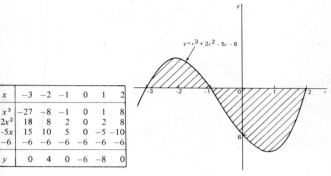

Figure 13.4

$$\int_{-3}^{-1} (x^3 + 2x^2 - 5x - 6)\ dx = \left[\frac{x^4}{4} + \frac{2x^3}{3} - \frac{5x^2}{2} - 6x\right]_{-3}^{-1}$$

$$= \left\{\frac{1}{4} - \frac{2}{3} - \frac{5}{2} + 6\right\} - \left\{\frac{81}{4} - 18 - \frac{45}{2} + 18\right\}$$

$$= \left\{3\frac{1}{12}\right\} - \left\{-2\frac{1}{4}\right\} = 5\frac{1}{3} \text{ square units}$$

$$\int_{-1}^{2} (x^3 + 2x^2 - 5x - 6)\ dx = \left[\frac{x^4}{4} + \frac{2x^3}{3} - \frac{5x^2}{2} - 6x\right]_{-1}^{2}$$

$$= \left\{4 + \frac{16}{3} - 10 - 12\right\} - \left\{3\frac{1}{12}\right\}$$

$$= \left\{-12\frac{2}{3}\right\} - \left\{3\frac{1}{12}\right\} = -15\frac{3}{4} \text{ square units}$$

Hence shaded area $= \left(5\frac{1}{3}\right) - \left(-15\frac{3}{4}\right) = \mathbf{21\frac{1}{12}}$ **square units**.

To determine the area enclosed between the curves $y = x^2 + 1$ and $y = 7 - x$:

At the points of intersection, the curves are equal. Thus, equating the y-values of each curve gives: $x^2 + 1 = 7 - x$, from which $x^2 + x - 6 = 0$. Factorising gives $(x - 2)(x + 3) = 0$, from which $x = 2$

199

and $x = -3$. By firstly determining the points of intersection the range of x-values have been found. Tables of values are produced as shown below

x	-3	-2	-1	0	1	2
$y = x^2 + 1$	10	5	2	1	2	5

x	-3	0	2
$y = 7 - x$	10	7	5

$y = 7 - x$ is a straight line thus only two points are needed, plus one more as a check. A sketch of the two curves is shown in *Figure 13.5*.

$$\text{Shaded area} = \int_{-3}^{2} (7 - x)\ dx - \int_{-3}^{2} (x^2 + 1)\ dx$$

$$= \int_{-3}^{2} [(7 - x) - (x^2 + 1)]\ dx = \int_{-3}^{2} (6 - x - x^2)\ dx$$

$$= \left[6x - \frac{x^2}{2} - \frac{x^3}{3} \right]_{-3}^{2} = \left(12 - 2 - \frac{8}{3} \right) - \left(-18 - \frac{9}{2} + 9 \right)$$

$$= \left(7\frac{1}{3} \right) - \left(-13\frac{1}{2} \right)$$

$$= 20\frac{5}{6} \text{ square units}.$$

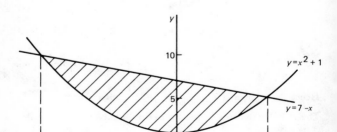

Figure 13.5

200

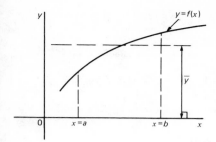

Figure 13.6

(b) Mean or average values

19   (i) The mean or average value of the curve shown in *Figure 13.6*, between $x = a$ and $x = b$ is given by:

$$\text{mean or average value, } \bar{y} = \frac{\text{area under curve}}{\text{length of base}}$$

(ii) When the area under a curve may be obtained by integration, then

$$\text{mean or average value, } \bar{y} = \frac{\int_a^b y \, dx}{b - a}$$

i.e., $y = \dfrac{1}{b-a} \displaystyle\int_a^b f(x) \, dx.$

(iii) For a periodic function, such as a sine wave, the mean value is assumed to be 'the mean value over half a cycle', since the mean value over a complete cycle is zero.

For example, the mean value of a voltage $v = 10 \sin \theta$ volts over half a cycle (i.e., between 0 and π rad) is given by:

$$\text{Mean value, } y = \frac{1}{\pi - 0} \int_0^\pi v \, d\theta = \frac{1}{\pi} \int_0^\pi 10 \sin \theta \, d\theta$$

$$=\frac{10}{\pi}[1-\cos\theta]_{0}^{\pi}=\frac{10}{\pi}\{(-\cos\pi)-(-\cos 0)\}$$

$$=\frac{10}{\pi}\{(1)-(-1)\}=\frac{20}{\pi}=\textbf{6.366 V}.$$

(For a sine wave, the mean value $=\frac{2}{\pi}\times$ maximum value, i.e.

$\frac{2}{\pi}\times 10=6.366$ V in this case.)

Root mean square value

20   The root mean square value of a quantity is 'the square root of the mean value of the squared values of the quantity' taken over an interval. With reference to *Figure 13.6*, the r.m.s. value of $y=f(x)$ over the range $x=a$ to $x=b$ is given by:

$$\textbf{r.m.s. value}=\sqrt{\left\{\frac{1}{b-a}\int_{a}^{b}y^{2}\ dx\right\}}$$

One of the principal applications of r.m.s. values is with alternating currents and voltages. The r.m.s. value of an alternating current is defined as that current which will give the same heating affect as the equivalent direct current.

For example, the r.m.s. value of a voltage $v=10\sin\theta$ volts over the range $\theta=0$ to $\theta=\pi$ is given by:

$$\text{r.m.s. value}=\sqrt{\left\{\frac{100}{\pi}\int_{0}^{\pi}v^{2}\ d\theta\right\}}=\sqrt{\left\{\frac{1}{\pi}\int_{0}^{\pi}(10\sin\theta)^{2}\ d\theta\right\}}$$

$$=\sqrt{\left\{\frac{100}{\pi}\int_{0}^{\pi}\sin^{2}\theta\ d\theta\right\}},\ \text{which is not a 'standard' integral.}$$

From 2 of *Table 13.1*, page 192, $\cos 2\theta=1-2\sin^{2}\theta$ and this formula is used whenever $\sin^{2}\theta$ needs to be integrated. Rearranging $\cos 2\theta=1-2\sin^{2}\theta$ gives $\sin^{2}\theta=\frac{1}{2}(1-\cos 2\theta)$. Hence

$$\sqrt{\left\{\frac{100}{\pi}\int_{0}^{\pi}\sin^{2}\theta\ d\theta\right\}}=\sqrt{\left\{\frac{100}{\pi}\int_{0}^{\pi}\frac{1}{2}(1-\cos 2\theta)d\theta\right\}}$$

$$= \sqrt{\left\{ \frac{100}{\pi} \left(\frac{1}{2} \right) \left[\theta - \frac{\sin 2\theta}{2} \right]_0^\pi \right\}}$$

$$= \sqrt{\left\{ \frac{100}{\pi} \left(\frac{1}{2} \right) \left[\left(\pi - \frac{\sin 2\pi}{2} \right) - \left(0 - \frac{\sin 0}{2} \right) \right] \right\}}$$

$$= \sqrt{\left\{ \frac{100}{\pi} \left(\frac{1}{2} \right) [\pi] \right\}} = \sqrt{\frac{100}{2}} = \frac{10}{\sqrt{2}}$$

$$= \mathbf{7.071 \ V}$$

(Note that for a sine wave, the r.m.s. value $= \dfrac{1}{\sqrt{2}} \times$ maximum value, i.e. $\dfrac{1}{\sqrt{2}} \times 10 = 7.071$ V in this case.)

(d) Volumes of solids of revolution

21 If the area under the curve $y = f(x)$, (shown in *Figure 13.7(a)*), between $x = a$ and $x = b$ is rotated 360° about the x-axis, then a volume known as a **solid of revolution** is produced as shown in *Figure 13.7(b)*. The volume of such a solid may be determined precisely using integration.

22   (i) Let the area shown in *Figure 13.7(a)* be divided into a number of strips each of width δx. One such strip is shown shaded.

    (ii) When the area is rotated 360° about the x-axis, each strip produces a solid of revolution approximating to a circular disc of radius y and thickness δx.

Volume of disc = (circular cross-sectional area) (thickness)
$= (\pi y^2)(\delta x)$.

    (iii) Total volume, V, between ordinates $x = a$ and $x = b$ is given by:

$$\textbf{Volume, } V = \lim_{\delta x \to 0} \sum_{x=a}^{x=b} \pi y^2 \delta x = \int_a^b \pi y^2 \, dx$$

Thus, for example, the volume of the solid of revolution produced when the curve $y = x^2 + 4$ is rotated one revolution about the x-axis between the limits $x = 1$ and $x = 4$ is given by:

$$\text{volume} = \int_1^4 \pi y^2 \, dx = \int_1^4 \pi (x^2 + 4)^2 \, dx$$

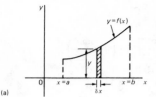

(a)

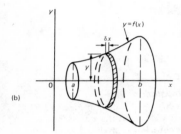

(b)

Figure 13.7

$$= \pi \int\limits_{1}^{4} (x^4 + 8x^2 + 16) \ dx$$

$$= \pi \left[\frac{x^5}{3} + \frac{8x^3}{3} + 16x \right]_{1}^{4}$$

$$= \pi [(204.8 + 170.67 + 64) - (0.2 + 2.67 + 16)]$$

$$= \textbf{420.6}\pi \ \textbf{cubic units}.$$

23   If a curve $x = f(y)$ is rotated about the y-axis 360° between the
limits $y = c$ and $y = d$, as shown in *Figure 13.8*, then the volume
generated is given by:

$$\textbf{Volume} = \mathop{\textbf{limit}}_{\boldsymbol{\delta y} \to \textbf{0}} \sum_{y=c}^{y=d} \pi x^2 \delta y = \int\limits_{c}^{d} \pi x^2 \ dy.$$

(e) Centroids

24   A **lamina** is a thin, flat sheet having uniform thickness. The
centre of gravity of a lamina is the point where it balances
perfectly, i.e. the lamina's **centre of mass**. When dealing with an

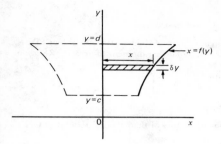

Figure 13.8

area (i.e. a lamina of negligible thickness and mass) the term
centre of area or **centroid** is used for the point where the centre of
gravity of a lamina of that shape would lie.

25　The first moment of area is defined as the product of the area
and the perpendicular distance of its centroid from a given axis in
the plane of the area. In *Figure 13.9*, the first moment of area A
about axis XX is given by (Ay) cubic units.

26　**Centroid of area between a curve and the x-axis**

(i) *Figure 13.10* shows an area PQRS bounded by the
curve $y = f(x)$, the x-axis and ordinates $x = a$ and $x = b$. Let
this area be divided into a large number of strips, each of
width δx.

A typical strip is shown shaded drawn at point (x, y)
on $f(x)$. The area of the strip is approximately rectangular
and is given by $y\delta x$.

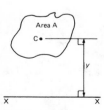

Figure 13.9

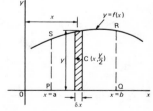

Figure 13.10

The centroid, C, has co-ordinates $\left(x, \dfrac{y}{2}\right)$

(ii) First moment of area of shaded strip about axis OY
$= (y\delta x)(x) = xy\delta x$. Total first moment of area PQRS about

axis OY $= \displaystyle\lim_{\delta x \to 0} \sum_{x=a}^{x=b} xy \,\delta x = \int_a^b xy \, dx.$

(iii) First moment of area of shaded strip about axis OX
$= (y\delta x)\left(\dfrac{y}{2}\right) = \dfrac{1}{2}y^2\delta x.$

Total first moment of area PQRS about axis OX
$= \displaystyle\lim_{\delta x \to 0} \sum_{x=a}^{x=b} \dfrac{1}{2} y^2\delta x.$

$= \dfrac{1}{2}\displaystyle\int_a^b y^2 \, dx$

(iv) Area of PQRS $A = \displaystyle\int_a^b y \, dx$ (from para 15).

(v) Let $\bar{x}$ and $\bar{y}$ be the distances of the centroid of area A about OY and OX respectively then:

$(\bar{x})(A) =$ total first moment of area A about axis OY $= \displaystyle\int_a^b xy \, dx$

from which, $\bar{x} = \dfrac{\displaystyle\int_a^b xy \, dx}{\displaystyle\int_a^b y \, dx},$

and $(\bar{y})(A) =$ total first moment of area A about axis OX

$= \dfrac{1}{2}\displaystyle\int_a^b y^2 \, dx$

$$\frac{1}{2}\int_a^b y^2\, dx$$

from which, $\bar{y} = \dfrac{\dfrac{1}{2}\displaystyle\int_a^b y^2\, dx}{\displaystyle\int_a^b y\, dx}$

Thus, to find the position of the centroid of the area bounded by the curve $y = 3x^2$, the x-axis and the ordinates $x = 0$ and $x = 2$:

$$\bar{x} = \frac{\displaystyle\int_0^2 xy\, dx}{\displaystyle\int_0^2 y\, dx} = \frac{\displaystyle\int_0^2 x(3x^2)\, dx}{\displaystyle\int_0^2 3x^2\, dx} = \frac{\displaystyle\int_0^2 3x^3\, dx}{\displaystyle\int_0^2 3x^2\, dx} = \frac{\left[\dfrac{3x^4}{4}\right]_0^2}{[x^3]_0^2} = \frac{12}{8} = \mathbf{1.5}$$

$$\bar{y} = \frac{\dfrac{1}{2}\displaystyle\int_0^2 y^2\, dx}{\displaystyle\int_0^2 y\, dx} = \frac{\dfrac{1}{2}\displaystyle\int_0^2 (3x^2)^2\, dx}{8} = \frac{\dfrac{1}{2}\displaystyle\int_0^2 9x^4\, dx}{8} = \frac{\dfrac{9}{2}\left[\dfrac{x^5}{2}\right]_0^2}{8} = \frac{\dfrac{9}{2}\left(\dfrac{32}{5}\right)}{8}$$

$$= \frac{18}{5} = \mathbf{3.6}$$

Hence the centroid lies at (1.5, 3.6).

27   Centroid of area between a curve and the y-axis
If $\bar{x}$ and $\bar{y}$ are the distances of the centroid
of area EFGH in *Figure 13.11* from OY
and OX respectively, then, by similar
reasoning as in para 26:

$$(\bar{x})(\text{total area}) = \lim_{\delta y \to 0} \sum_{y=c}^{y=d} (x\, \delta y)\left(\frac{x}{2}\right) = \frac{1}{2}\int_c^d x^2\, dy,$$

from which, $\quad \bar{x} = \dfrac{\frac{1}{2}\int_c^d x^2\, dy}{\int_c^d x\, dy}$

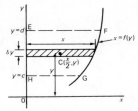

Figure 13.11

and $(\bar{y})$ (total area) $= \displaystyle\lim_{\delta y \to 0} \int_{y=c}^{y=d} (x\delta y)y = \int_c^d xy\, dy.$

from which, $\quad \bar{y} = \dfrac{\int_c^d xy\, dy}{\int_c^d x\, dy}$

(f) Second moments of areas

28   (i) The **first moment of area** about a fixed axis of a lamina of area A, perpendicular distance y from the centroid of the lamina is defined as Ay cubic units.

    (ii) The **second moment of area** of the same lamina as in (i) is given by Ay^2, i.e., the perpendicular distance from the centroid of the area to the fixed axis is squared.

    (iii) Second moments of areas are usually denoted by I and have units of mm^4, cm^4, and so on.

29   Several areas, a_1, a_2, a_3, ... at distances y_1, y_2, y_3, ... from a fixed axis, may be replaced by a single area A, where $A = a_1 + a_2 + a_3 + \ldots$ at distance k from the axis, such that $Ak^2 = \sum ay^2$. k is called the radius of gyration of Area A about the given axis. Since $Ak^2 = \sum ay^2 = I$ then the **radius of gyration**, $k = \sqrt{\left(\dfrac{I}{A}\right)}$.

30   The second moment of area is a quantity much used in the theory of bending of beams, in the torsion of shafts, and in calculations involving water planes and centres of pressure.

31   The procedure to determine the second moment of area of regular sections about a given axis is (i) to find the second moment of area of a typical element and (ii) to sum all such second moments of area by integrating between appropriate limits.

    For example, **the second moment of area of the rectangle** shown in *Figure 13.12* about axis PP is found by initially

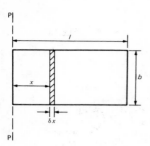

Figure 13.12

considering an elemental strip of width δ_x and parallel to and distance x from axis PP. Area of shaded strip $= b\delta x$.

Second moment of area of the shaded strip about PP $= (x^2)(b\ \delta x)$. The second moment of area of the whole rectangle about PP is obtained by summing all such strips between $x = 0$ and $x = l$, i.e.

$$\sum_{x=0}^{x=l} x^2 b\delta x.$$

It is a fundamental theorem of integration that

$$\lim_{\delta x \to 0} \sum_{x=0}^{x=l} x^2 b\delta x = \int_0^l x^2 b\ dx.$$

Thus the second moment of area of the rectangle about

$$PP = b \int_0^l x^2\ dx$$

$$= b\left[\frac{x^3}{3}\right]_0^l = \frac{bl^3}{3}$$

Since the total area of the rectangle,

$$A = lb,\ \text{then}\ I_{PP} = (lb)\left(\frac{l^2}{3}\right) = \frac{Al^2}{3}$$

$$I_{PP} = Ak_{PP}^2\ \text{thus}\ k_{PP}^2 = \frac{l^2}{3}$$

209

i.e. the radius of gyration about axis PP,

$$k_{PP} = \sqrt{\left(\frac{l^2}{3}\right)} = \frac{l}{\sqrt{3}}$$

A polar axis is an axis through the centre of a circle, perpendicular to the plane of the circle as shown by axis ZZ in *Figure 13.13*. Consider the elemental annulus of width δx shown shaded in *Figure 13.13*, radius x from the centre, 0. If δx is very small, then the area of the elemental annulus is approximately given by (circumference × width), i.e. area of annulus $\approx (2\pi x)(\delta x)$.

Second moment of area of annulus about the centre of the circle $\approx x^2(2\pi r\,dx)$. Total second moment of area of the circle about ZZ,

$$l_{ZZ} = \underset{\delta x \to 0}{\text{limit}} \sum_{x=0}^{x=r} x^2(2\pi x\delta x)$$

i.e. $l_{ZZ} = \int\limits_0^r 2\pi x^3\,dx = 2\pi\left[\frac{x^4}{4}\right]_0^r = \frac{\pi r^4}{2}$

Radius of gyration, $k_{ZZ} = \sqrt{\frac{I_{ZZ}}{\text{area}}} = \sqrt{\left(\frac{\frac{\pi r^4}{2}}{\pi r^2}\right)} = \frac{r}{\sqrt{2}}$

Parallel axis theorem

32   In *Figure 13.14*, axis GG passes through the centroid C of area A. Axes DD and GG are in the same plane, are parallel to each other and distance d apart. The parallel axis theorem states:

$$I_{DD} = I_{GG} + Ad^2$$

Using the parallel axis theorem the second moment of area of a rectangle about an axis through the centroid may be determined.

In the rectangle shown in *Figure 13.15*, $I_{PP} = \dfrac{bl^3}{3}$ (from para 31).

From the parallel axis theorem

$$I_{PP} = I_{GG} + (bl)\left(\frac{l}{2}\right)^2$$

i.e. $\dfrac{bl^3}{3} = I_{GG} + \dfrac{bl^3}{4}$

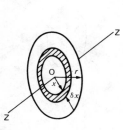

Figure 13.13

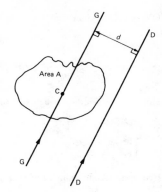

Figure 13.14

from which,

$$I_{GG} = \frac{bl^3}{3} - \frac{bl^3}{4} = \frac{bl^3}{12}$$

Perpendicular axis theorem

33 In *Figure 13.16*, axes OX, OY and OZ are mutually perpendicular. If OX and OY lie in the plane of area A then the perpendicular axis theorem states:

$$I_{OZ} = I_{OX} + I_{OY}$$

A circle of radius r, having three mutually perpendicular axes OX,

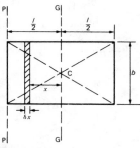

Figure 13.15

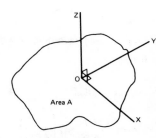

Figure 13.16

OY and ZZ is shown in *Figure 13.17*. From para 31,

$$I_{ZZ} = \frac{\pi r^4}{2}$$

By symmetry,

$$I_{OX} = I_{OY}$$

Using the perpendicular axis theorem:

$$I_{ZZ} = I_{OX} + I_{OY} = 2I_{OX}$$

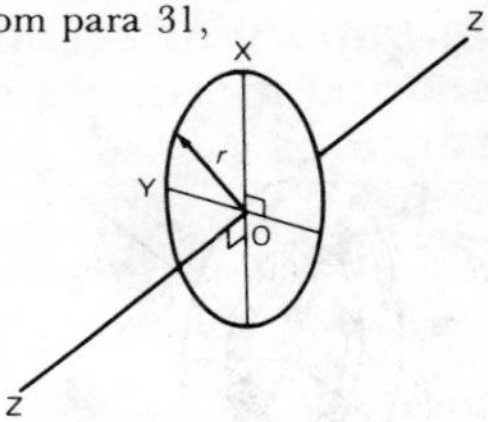

Figure 13.17

i.e. $\dfrac{\pi r^4}{2} = 2I_{OX}$, from which, $I_{OX} = \dfrac{\pi r^4}{4}$

Hence the second moment of area of a circle about a diameter is $\dfrac{\pi r^4}{4}$.

Radius of gyration, $k_{OX} = \sqrt{\dfrac{I_{OX}}{\text{area}}} = \sqrt{\left(\dfrac{\dfrac{\pi r^4}{4}}{\pi r^2}\right)} = \dfrac{r}{2}$.

Figure 13.18 shows a circle of radius r having a tangent PP parallel to diameter XX. By the parallel axis theorem:

$$I_{PP} = I_{XX} + Ar^2$$

From above,

$$I_{XX} = \frac{\pi r^4}{4}.$$

Thus

$$I_{PP} = \frac{\pi r^4}{4} (\pi r^2) r^2 = \frac{5}{4} \pi r^4.$$

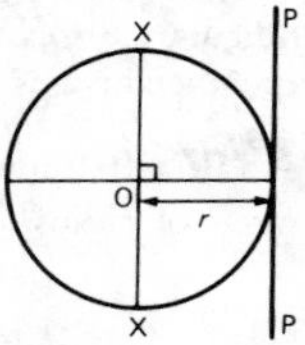

Figure 13.18

Thus the second moment of area of a circle about a tangent is $\dfrac{5\pi}{4} r^4$.

Radius of gyration,

$$k_{PP} = \sqrt{\dfrac{I_{PP}}{\text{area}}} = \sqrt{\left(\dfrac{\dfrac{5}{4}\pi r^4}{\pi r^2}\right)} = \dfrac{\sqrt{5}}{2} r.$$

Summary of standard results of the second moments of areas of regular sections

Shape	Position of axis	Second moment of area, I	Radius of gyration, k
Rectangle length l breadth b	(1) Coinciding with b	$\dfrac{bl^3}{3}$	$\dfrac{l}{\sqrt{3}}$
	(2) Coinciding with l	$\dfrac{lb^3}{3}$	$\dfrac{b}{\sqrt{3}}$
	(3) Through centroid, parallel to b.	$\dfrac{bl^3}{12}$	$\dfrac{l}{\sqrt{12}}$
	(4) Through centroid, parallel to l.	$\dfrac{lb^3}{12}$	$\dfrac{b}{\sqrt{12}}$
Triangle Perpendicular height h base b	(1) Coinciding with b	$\dfrac{bh^3}{12}$	$\dfrac{h}{\sqrt{6}}$
	(2) Through centroid, parallel to base	$\dfrac{bh^3}{36}$	$\dfrac{h}{\sqrt{18}}$
	(3) Through vertex parallel to base	$\dfrac{bh^3}{4}$	$\dfrac{h}{\sqrt{2}}$
Circle radius r	(1) Through centre, perpendicular to plane (i.e. polar axis)	$\dfrac{\pi r^4}{2}$	$\dfrac{r}{\sqrt{2}}$
	(2) Coinciding with diameter	$\dfrac{\pi r^4}{4}$	$\dfrac{r}{2}$
	(3) About a tangent	$\dfrac{5\pi}{4}r^4$	$\dfrac{\sqrt{5}}{2}r$
Semicircle radius r	Coinciding with diameter	$\dfrac{\pi r^4}{8}$	$\dfrac{r}{2}$

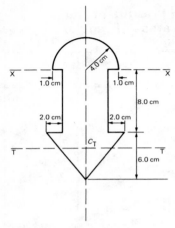

Figure 13.19

For example, to determine the second moment of area about axis
XX for the composite area shown in *Figure 13.19*:

For the semicircle, $I_{XX} = \dfrac{\pi r^4}{8} = \dfrac{\pi (4.0)^4}{8} = 100.5$ cm^4

For the rectangle, $I_{XX} = \dfrac{bl^3}{3} = \dfrac{(6.0)(8.0)^3}{3} = 1024$ cm^4

For the triangle, about axis TT through centroid C_T,

$$I_{TT} = \frac{bh^3}{36} = \frac{(10.0)(6.0)^3}{36} = 60 \text{ cm}^4.$$

By the parallel axis theorem, the second moment of area of the
triangle about axis XX.

$$= 60 + \left[\frac{1}{2}(10.0)(6.0)\right]\left[8.0 + \frac{1}{3}(6.0)\right] = 360 \text{ cm}^4.$$

Total second moment of area about XX $= 100.5 + 1024 + 360$

$= 1484.5 = $ **1480 cm^4**, correct to 3 significant figures.

14 Differential equations

1   A **differential equation** is one that contains differential

coefficients. Examples include: (i) $\dfrac{dy}{dx} = 7x$ and (ii) $\dfrac{d^2y}{dx^2} + 5\dfrac{dy}{dx} + 2y = 0$

2   Differential equations are classified according to the highest
derivative which occurs in them. Thus example (i) above is a **first
order differential equation**, and example (ii) is a **second order
differential equation**.

3   Starting with a differential equation it is possible, by
integration and by being given sufficient data to determine
unknown constants, to obtain the original function. This process is
called **'solving the differential equation'**.

4   (i) A solution to a differential equation which contains
        one or more arbitrary constants of integration is called the
        general solution of the differential equation.
        (ii) When additional information is given so that
        constants may be calculated the **particular solution** of
        the differential equation is obtained. The additional
        information is called **boundary conditions**.

5   Differential equations are widely used in engineering and
science. There are several different type of differential equation
and each requires their own method of solution.

(a) $\dfrac{dy}{dx} = f(x)$ type

6 . A differential equation of the form $\dfrac{dy}{dx} = f(x)$ is solved by direct

integration, i.e.

$$y = f(x)\ dx.$$

For example, to solve the differential equation

$\dfrac{dy}{dx} = 3x^2 - \sin 2x$, given the boundary conditions $y = 1\dfrac{1}{2}$ when $x = 0$:

Integrating both sides of $\dfrac{dy}{dx} = 3x^2 - \sin 2x$ gives:

$$\int \frac{dy}{dx} dx = \int (3x^2 - \sin 2x) \ dx$$

i.e. $y = x^3 + \dfrac{1}{2} \cos 2x + c$, which is the general solution.

$y = 1\dfrac{1}{2}$ when $x = 0$, hence $1\dfrac{1}{2} = 0 + \dfrac{1}{2} \cos 0 + c$

from which, $c = 1$.

Thus the particular solution is:

$y = x^3 + \dfrac{1}{2} \cos 2x + 1.$

(b) $\dfrac{dy}{dx} = f(y)$ type

7   A differential equation of the form $\dfrac{dy}{dx} = f(y)$ is initially

rearranged to give $dx = \dfrac{dy}{f(y)}$, and then the solution is obtained by

direct integration,

i.e. $\displaystyle\int dx = \int \frac{dy}{f(y)}$

Thus, for example, to determine the particular solution of

$$(y^2 - 1) \frac{dy}{dx} = 3y, \text{ given } y = 1 \text{ when } x = 2\frac{1}{6},$$

the differential equation is firstly rearranged giving:

$$dx = \frac{(y^2 - 1)}{3y} \ dy = \left(\frac{y}{3} - \frac{1}{3y} \right) dy$$

Integrating gives:

$$\int dx = \int \left(\frac{y}{3} - \frac{1}{3y} \right) dy$$

i.e. $x = \dfrac{y^2}{6} - \dfrac{1}{3} \ln y + c$, which is the general solution.

$y = 1$ when $x = 2\dfrac{1}{6}$, hence $2\dfrac{1}{6} = \dfrac{1}{6} - \dfrac{1}{3} \ln 1 + c$

from which, $c = 2$.

Hence the particular solution is:

$$x = \frac{y^2}{6} - \frac{1}{3} \ln y + 2.$$

(c) $\dfrac{dy}{dx} = f(x).g(y)$

8   A differential equation of the form $\dfrac{dy}{dx} = f(x).g(y)$, where $f(x)$ is a function of x only and $g(y)$ is a function of y only, may be rearranged as $\dfrac{dy}{g(y)} = f(x)\ dx$, and then the solution is obtained by direct integration, i.e.

$$\int \frac{dy}{g(y)} = \int f(x)\ dx.$$

When two variables are rearranged into two separate groups as shown above, each containing only one variable, the variables are said to be separable.

Differential equations of the forms $\dfrac{dy}{dx} = f(x)$ and $\dfrac{dy}{dx} = f(y)$ are merely special cases of **'separating the variables'**.

For example, to solve $\dfrac{dy}{dx} = 2y \cos x$, given $y = 1$ when $x = \dfrac{\pi}{2}$, the variables are first separated giving: $\dfrac{dy}{2y} = \cos x\ dx$.

Integrating both sides gives:

$$\int \frac{dy}{2y} = \int \cos x\ dx$$

i.e. $\dfrac{1}{2} \ln y = \sin x + c$

$y = 1$ when $x = \dfrac{\pi}{2}$, hence $\dfrac{1}{2} \ln 1 = \sin \dfrac{\pi}{2} + c$ from which $c = -1$

Hence the solution is $\dfrac{1}{2} \ln y = \sin x - 1$

(d) $\dfrac{dQ}{dt} = kQ$ type

9   The general solution of an equation of the form $\dfrac{dQ}{dt} = kQ$ is $Q = Ae^{kt}$ where A is a constant.

This solution may be checked: Differentiating $Q = Ae^{kt}$ with respect to t gives: $\dfrac{dQ}{dt} = k(Ae^{kt}) = kQ.$

This provides an alternative method of solution of type (b) given in para 7.

For example, in solving $\dfrac{dy}{dx} = 5y$, given $y = 2$ when $x = 1$, it is

recognised that $\dfrac{dy}{dx} = 5y$ is of the form $\dfrac{dQ}{dt} = kQ$ (where $y = Q$, $x = t$

and $k = 5$).

Hence the solution is $y = Ae^{5x}$.

Given $y = 2$ when $x = 1$ enables A to be determined.

Thus $2 = Ae^{5(1)}$, i.e., $A = \dfrac{2}{e^5} = 2e^{-5}$.

Hence the particular solution is:

$y = (2e^{-5})e^{5x}$

i.e. $y = 2e^{5(x-1)}$

which may also be obtained using the method shown in (b).

10   Examples of the natural laws of the form $\dfrac{dQ}{dt} = kQ$ include:

     (i)   **Newton's law of cooling:** $\dfrac{d\theta}{dt} = -k\theta.$ The law is $\theta = \theta_0 e^{-kt}$

     (ii)   **Decay of current in an inductive circuit:** $\dfrac{di}{dt} = ki.$ The

     law is $i = Ae^{kt}$

     (iii)   **Linear expansion:** $\dfrac{dl}{d\theta} = kl.$ The law is $l = l_0 e^{k\theta}$

(e) Homogeneous first order differential equation

11   Certain first order differential equations are not of the 'variable-separable' type but can be made separable by changing the variable.

12   An equation of the form $P\dfrac{dy}{dx} = Q$, where P and Q are

functions of both x and y of the same degree throughout, is said to be **homogeneous** in y and x.

     For example, $f(x, y) = x^2 + 3xy + y^2$ is a homogeneous function since each of the three terms are of degree 2. Similarly,

218

$f(x, y) = \dfrac{x - 3y}{2x + y}$ is homogeneous in x and y since each of the four terms are degree 1.

However, $f(x, y) = \dfrac{x^2 - y}{2x^2 + y^2}$ is not homogeneous since the term in y in the numerator is of degree 1 and the other three terms are of degree 2.

Procedure to solve differential equations of the form $P\dfrac{dy}{dx} = Q$

13 (i) Rearrange $P\dfrac{dy}{dx} = Q$ into the form $\dfrac{dy}{dx} = \dfrac{Q}{P}$.

(ii) Make the substitution $y = vx$ (where v is a function of x), from which, $\dfrac{dy}{dx} = v(1) + x\dfrac{dv}{dx}$, by the product rule.

(iii) Substitute for both y and $\dfrac{dy}{dx}$ in the equation $\dfrac{dy}{dx} = \dfrac{Q}{P}$. Simplify by cancelling, and an equation results in which the variables are separable.

(iv) Separate the variables and solve using the method shown in para 8.

(v) Substitute $v = \dfrac{y}{x}$ to solve in terms of the original variables.

For example, to solve the equation $x\dfrac{dy}{dx} = \dfrac{x^2 + y^2}{y}$, given that $x = 1$ when $y = 4$, using the above procedure:

(i) Rearranging $x\dfrac{dy}{dx} = \dfrac{x^2 + y^2}{y}$ gives $\dfrac{dy}{dx} = \dfrac{x^2 + y^2}{xy}$ which is homogeneous in x and y since each of the three terms on the right hand side are of the same degree (i.e. degree 2).

(ii) Let $y = vx$ then $\dfrac{dy}{dx} = v + x\dfrac{dv}{dx}$

(iii) Substituting for y and $\dfrac{dy}{dx}$ in the equation $\dfrac{dy}{dx} = \dfrac{x^2 + y^2}{xy}$ gives:

$$v + x\dfrac{dv}{dx} = \dfrac{x^2 + (vx)^2}{x(vx)} = \dfrac{x^2 + v^2 x^2}{vx^2} = \dfrac{1 + v^2}{v}$$

219

(iv) Separating the variables gives: $x \dfrac{dv}{dx} = \dfrac{1 + v^2}{v} - v$

$$= \frac{1 + v^2 - v^2}{v} = \frac{1}{v}$$

Hence, $v \, dv = \dfrac{1}{x} \, dx$.

Integrating both sides gives: $\displaystyle\int v \, dv = \int \frac{1}{x} \, dx$, i.e. $\dfrac{v^2}{2} = \ln x + c$

(v) Replacing v by $\dfrac{y}{x}$ gives: $\dfrac{y^2}{2x^2} = \ln x + c$, which is the general solution.

When $x = 1$, $y = 4$, thus: $\dfrac{16}{2} = \ln 1 + c$ from which, $c = 8$.

Hence the particular solution is $\dfrac{y^2}{2x^2} = \ln x + 8$

or $y^2 = 2x^2(\ln x + 8)$.

(f) Linear first order differential equations

14   An equation of the form $\dfrac{dy}{dx} + Py = Q$, where P and Q are

functions of x only is called a **linear differential equation** since y and its derivatives are of the first degree.

15   (i) The solution of $\dfrac{dy}{dx} + Py = Q$ is obtained by multiplying

throughout by what is termed an **integrating factor**.

(ii) Multiplying $\dfrac{dy}{dx} + Py = Q$ by say R, a function of x

only, gives

$$R\frac{dy}{dx} + RPy = RQ \qquad (1)$$

(iii) The differential coefficient of a product Ry is

obtained using the product rule, i.e. $\dfrac{d}{dx}(Ry) = R\dfrac{dy}{dx} + y\dfrac{dR}{dx}$,

which is the same as the left-hand side of equation (1),

when R is such that $RP = \dfrac{dR}{dx}$.

(iv) If $\dfrac{dR}{dx} = RP$, then separating the variables gives

$\dfrac{dR}{R} = P\ dx$. Integrating both sides gives $\displaystyle\int \dfrac{dR}{R} = \int P\ dx$

i.e. $\ln R = \int P\ dx + c$

from which, $R = e^{\int P\ dx + c} = e^{\int P\ dx} e^{c}$

i.e. $R = Ae^{\int P\ dx}$, where $A = e^{c} = $ a constant

(v) Substituting $R = Ae^{\int P\ dx}$ in equation (1) gives:

$$Ae^{\int P\ dx}\left(\dfrac{dy}{dx}\right) + Ae^{\int P\ dx}P\,y = Ae^{\int P\ dx}\,Q$$

i.e., $\ e^{\int P\ dx}\left(\dfrac{dy}{dx}\right) + e^{\int P\ dx}P\,y = e^{\int P\ dx}Q$ \hfill (2)

(vi) The left-hand side of equation (2) is $\dfrac{d}{dx}(y e^{\int P\ dx})$,

which may be checked by differentiating $y e^{\int P\ dx}$ with respect to x, using the product rule.

(vii) From equation (2), $\dfrac{d}{dx}(y e^{\int P\ dx}) = e^{\int P\ dx}Q$ Integrating

both sides gives:

$$y e^{\int P\ dx} = \int e^{\int P\ dx}Q\ dx$$ \hfill (3)

(viii) $e^{\int P\ dx}$ is the integrating factor.

Procedures to solve differential equations of the form $\dfrac{dy}{dx} + Py = Q$

16   (i) Rearrange the differential equation into the form
$\dfrac{dy}{dx} + Py = Q$, where P and Q are functions of x.

(ii) Determine $\int P\ dx$.

(iii) Determine the integrating factor $e^{\int P\ dx}$

(iv) Substitute $e^{\int P\ dx}$ into equation (3).

(v) Integrate the right-hand side of equation (3) to give the general solution of the differential equation. Given boundary conditions, the particular solution may be determined.

For example, to solve $\dfrac{1}{x}\dfrac{dy}{dx} + 4y = 2$ given $x = 0$ when $y = 4$,

221

using the above procedure:

(a) Rearranging gives $\dfrac{dy}{dx} + 4xy = 2x$, which is of the form

$\dfrac{dy}{dx} + Py = Q$, where $P = 4x$ and $Q = 2x$.

(b) $\int P \, dx = \int 4x \, dx = 2x^2$

(c) Integrating factor $e^{\int P \, dx} = e^{2x^2}$

(d) Substitution into equation (3) gives: $ye^{2x^2} = \int e^{2x^2}(2x) \, dx$

(e) Hence the general solution is: $ye^{2x^2} = \dfrac{1}{2}e^{2x^2} + c$, by using the substitution $u = 2x^2$.

When $x = 0$, $y = 4$, thus $4e^0 = \dfrac{1}{2}e^0 + c$, from which, $c = \dfrac{7}{2}$

Hence the particular solution is $ye^{2x^2} = \dfrac{1}{2}e^{2x^2} + \dfrac{7}{2}$ or

$y = \dfrac{1}{2} + \dfrac{7}{2}e^{-2x^2}$ or $y = \dfrac{1}{2}(1 + 7e^{-2x^2})$.

(g) $a\dfrac{d^2y}{dx^2} + b\dfrac{dy}{dx} + cy = 0$ type

17 An equation of the form $a\dfrac{d^2y}{dx^2} + b\dfrac{dy}{dx} + cy = 0$, where a, b and c

are constants, is called a **linear second order differential equation with constant coefficients**.

18 If D represents $\dfrac{d}{dx}$ and D^2 represents $\dfrac{d^2}{dx^2}$, then the equation in para 17 may be stated as $(aD^2 + bD + c)y = 0$. This equation is said to be in 'D-operator' form.

19 If $y = Ae^{mx}$, then $\dfrac{dy}{dx} = Ame^{mx}$ and $\dfrac{d^2y}{dx^2} = Am^2e^{mx}$.

Substituting these values into $a\dfrac{d^2y}{dx^2} + b\dfrac{dy}{dx} + cy = 0$ gives:

$a(Am^2e^{mx}) + b(Ame^{mx}) + c(Ae^{mx}) = 0$

i.e., $Ae^{mx}(am^2 + bm + c) = 0$.

Thus $y = Ae^{mx}$ is a solution of the given equation provided that $(am^2 + bm + c) = 0$.

20 $am^2 + bm + c = 0$ is called the **auxiliary equation**, and since the equation is a quadratic, m may be obtained either by factorising or by using the quadratic formula.

21 Since, in the auxiliary equation, a, b and c are real values,

then the equation may have either

      (i) two different real roots (when $b^2 > 4ac$), or

      (ii) two real equal roots (when $b^2 = 4ac$), or

      (iii) two complex roots (when $b^2 < 4ac$)

Procedure to solve differential equations of the form $a\dfrac{d^2y}{dx^2} + b\dfrac{dy}{dx} + cy = 0$

22 (a) Rewrite the differential equation $a\dfrac{d^2y}{dx^2} + b\dfrac{dy}{dx} + cy = 0$ as

$(aD^2 + bD + c)y = 0$.

(b) Substitute m for D, and solve the auxiliary equation $am^2 + bm + c = 0$, for m.

(c) If the roots of the auxiliary equation are:

    (i) **real and different**, say $m = \alpha$ and $m = \beta$, then the general solution is

$y = Ae^{\alpha x} + Be^{\beta x}$

    (ii) **real and equal**, say $m = \alpha$ twice, then the general solution is

$y = (Ax + B)e^{\alpha x}$,

    (iii) **complex**, say $m = \alpha \pm j\beta$, then the general solution is
$y = e^{\alpha x}\{C \cos \beta x + D \sin \beta x\}$.

(d) Given boundary conditions, constants A and B, or C and D, may be determined and the **particular solution** of the differential equation obtained.

For example, to solve $2\dfrac{d^2y}{dx^2} + 5\dfrac{dy}{dx} - 3y = 0$, given that when

$x = 0$, $y = 4$ and $\dfrac{dy}{dx} = 9$, using the above procedure:

(i) $2\dfrac{d^2y}{dx^2} + 5\dfrac{dy}{dx} - 3y = 0$ in D-operator form is

$(2D^2 + 5D - 3)y = 0$, where $D \equiv \dfrac{d}{dx}$.

(ii) Substituting m for D gives the auxiliary equation $2m^2 + 5m - 3 = 0$

Factorising gives: $(2m - 1)(m + 3) = 0$, from which $m = \dfrac{1}{2}$ or

$m = -3$

(iii) Since the roots are real and different the **general solution** is $y = Ae^{(1/2)x} + Be^{-3x}$

(iv) When $x=0$, $y=4$, hence $4 = A + B$ (1)

Since $y = Ae^{(1/2)x} + Be^{-3x}$ then $\dfrac{dy}{dx} = \dfrac{1}{2}Ae^{(1/2)x} - 3Be^{-3x}$

When $x=0$, $\dfrac{dy}{dx} = 9$, thus $9 = \dfrac{1}{2}A - 3B$ (2)

Solving the simultaneous equations (1) and (2) gives $A = 6$ and $B = -2$. **Hence the particular solution is $y = be^{(1/2)}x - 3e^{-3x}$**

To solve $9\dfrac{d^2y}{dt^2} - 24\dfrac{dy}{dx} + 16y = 0$, given that when $t=0$,

$y = \dfrac{dy}{dt} = 3$, using the above procedure:

(a) $9\dfrac{d^2y}{dt^2} - 24\dfrac{dy}{dt} + 16y = 0$ in D-operator form is

$(9D^2 - 24D + 16)y = 0$, where $D \equiv \dfrac{d}{dt}$.

(b) Substituting m for D gives the auxiliary equation $9m^2 - 24m + 16 = 0$. Factorising gives: $(3m-4)(3m-4) = 0$, i.e., $m = \dfrac{4}{3}$ twice.

(c) Since the roots are real and equal, the **general solution** is

$$y = (At + B)e^{(4/3)t}.$$

(d) When $t=0$, $y=3$ hence $3 = (0+B)e^0$, i.e., $B = 3$.

Since $y = (At+B)e^{(4/3)t}$, then $\dfrac{dy}{dt} = (At+B)\left(\dfrac{4}{3}e^{(4/3)t}\right) + Ae^{(4/3)t}$,

by the product rule.

When $t=0$, $\dfrac{dy}{dt} = 3$ thus $3 = (0+B)\dfrac{4}{3}e^0 + Ae^0$,

i.e. $3 = \dfrac{4}{3}B + A$

from which, $A = -1$, since $B = 3$.

Hence the particular solution is

$$y = (+t+3)e^{(4/3)t} \text{ or } y = (3-t)e^{\frac{4}{3}t}.$$

To solve $\dfrac{d^2y}{dx^2} + 6\dfrac{dy}{dx} + 13y = 0$, given that when $x=0$, $y=3$

224

and $\dfrac{dy}{dx} = 7$, using the above procedure:

(a) $\dfrac{d^2y}{dx^2} + 6\dfrac{dy}{dx} + 13y = 0$ in D-operator form is $(D^2 + 6D + 13)y = 0$

where $D \equiv \dfrac{d}{dx}$.

(b) Substituting m for D gives the auxiliary equation
$m^2 + 6m + 13 = 0$.
Using the quadratic formula: m

$$m = \frac{-6 \pm \sqrt{[(6)^2 - 4(1)(13)]}}{2(1)} = \frac{-6 \pm \sqrt{(-16)}}{2}$$

i.e. $m = \dfrac{-6 \pm j4}{2} = -3 \pm j2$

(c) Since the roots are complex, the **general solution is**
$y = e^{-3x}(C \cos 2x + D \sin 2x)$.

(d) When $x = 0$, $y = 3$ hence $3 = e^0 \ (C \cos 0 + D \sin 0)$, i.e.
$C = 3$.
Since $y = e^{-3x}(C \cos 2x + D \sin 2x)$,
Then $\dfrac{dy}{dx} = e^{-3x} \ (-2C \sin 2x + 2D \cos 2x) -$

$3e^{-3x}(C \cos 2x + D \sin 2x)$, by the product rule,
$= e^{-3x}\{(2D - 3C) \cos 2x - (2C + 3D) \sin 2x\}$

When $x = 0$, $\dfrac{dy}{dx} = 7$, hence

$7 = e^0\{(2D - 3C) \cos 0 - (2C + 3D) \sin 0\}$
i.e., $7 = 2D - 3C$, from which, $D = 8$, since $C = 3$
Hence the particular solution is
$y = e^{-3x}(3 \cos 2x + 8 \sin 2x)$
Since $a \cos \omega t + b \sin \omega t = R \sin(\omega t + \alpha)$, where

$R = \sqrt{(a^2 + b^2)}$ and $\alpha = \arctan \dfrac{a}{b}$, (see para 26, page 150), then

$3 \cos 2x + 8 \sin 2x = \sqrt{(3^2 + 8^2)} \sin\left(2x + \arctan\dfrac{3}{8}\right)$
$= \sqrt{73} \sin(2x + 20° \ 33') = \sqrt{73} \sin(2x + 0.359)$
Thus the particular solution may also be expressed as
$y = \sqrt{73}e^{-3x}\sin(2x + 0.359)$

(h) $a\dfrac{d^2y}{dx^2} + b\dfrac{dy}{dx} + cy = f(x)$ type

23   If in the differential equation $a\dfrac{d^2y}{dx^2} + b\dfrac{dy}{dx} + cy = f(x)$ (1)

the substitution $y = u + v$ is made then:

$$a\frac{d^2(u+v)}{dx^2} + b\frac{d(u+v)}{dx} + c(u+v) = f(x)$$

Rearranging gives:

$$\left(a\frac{d^2u}{dx^2} + b\frac{du}{dx} + cu\right) + \left(a\frac{d^2v}{dx^2} + b\frac{dv}{dx} + cv\right) = f(x)$$

If we let

$$a\frac{d^2v}{dx^2} + b\frac{dv}{dx} + cv = f(x) \qquad (2)$$

then

$$a\frac{d^2u}{dx^2} + b\frac{du}{dx} + cu = 0 \qquad (3)$$

24   The general solution, u, of equation (3) will contain two unknown constants, as required for the general solution of equation (1). The method of solution of equation (3) is shown in para 22. The function u is called the **complementary function, (CF).**

25   If the particular solution, v, of equation (2) can be determined without containing any unknown constants then $y = u + v$ will give the general solution of equation (1). The function v is called the **particular integral, (PI)**. Hence the general solution of equation (1) is given by:

$$y = \mathbf{CF} + \mathbf{PI}$$

Procedure to solve differential equations of the form $a\dfrac{d^2y}{dx^2} + b\dfrac{dy}{dx} + cy = f(x)$

26   (i) Rewrite the given differential equation as $(aD^2 + bD + c)y = f(x)$.

(ii) Substitute m for D, and solve the auxiliary equation

$am^2 + bm + c = 0$ for m.

(iii) Obtain the complementary function, u, which is achieved using the same procedure as in para 22.

226

(iv) To determine the particular integral, v, firstly assume a particular integral which is suggested by $f(x)$, but which contains undetermined coefficients. *Table 14.1* gives some suggested substitutions for different functions $f(x)$.

(v) Substitute the suggested PI into the differential equation

$$(aD^2 + bD + c) = f(x)$$

and equate relevant coefficients to find the constants introduced.

(vi) The general solution is given by $y = CF + PI$, i.e. $y = u + v$.

(vii) Given boundary conditions, arbitrary constants in the CF may be determined and the particular solution of the differential equation obtained.

For example, to solve $\dfrac{d^2y}{dx^2} - 3\dfrac{dy}{dx} = 9$, given that when $x = 0$, $y = 0$

and $\dfrac{dy}{dx} = 0$, using the procedure for para 26:

(i) $\dfrac{d^2y}{dx^2} - 3\dfrac{dy}{dx} = 9$ in D-operator form is $(D^2 - 3D)y = 9$

(ii) Substituting m for D gives the auxiliary equation $m^2 - 3m = 0$

Factorising gives: $m(m - 3) = 0$, from which, $m = 0$ or $m = 3$

(iii) Since the roots are real and different, the CF,

$$u = Ae^0 + Be^{3x}, \text{ i.e., } \boldsymbol{u = A + Be^{3x}}$$

(iv) Since the CF contains a constant (i.e. A) then let the PI, $v = kx$ (see *Table 14.1(a)*).

(v) Substituting $v = kx$ into $(D^3 - 3D)v = 9$ gives $(D^2 - 3D)kx = 9$, $D(kx) = k$ and $D^2(kx) = 0$

Hence $(D^2 - 3D)kx = 0 - 3k = 9$, from which, $k = -3$

Hence the PI, $v = -3x$

(vi) The general solution is given by $y = u + v$, i.e.

$$\boldsymbol{y = A + Be^{3x} - 3x}$$

(vii) When $x = 0$, $y = 0$ then $0 = A + Be^0 - 0$, i.e. $0 = A + B$ (1)

$\dfrac{dy}{dx} = 3Be^{3x} - 3$; $\dfrac{dy}{dx} = 0$ when $x = 0$, thus $0 = 3Be^0 - 3$, from

which, $B = 1$. From equation (1), $A = -1$

Hence the particular solution is $y = -1 + 1e^{3x} - 3x$, i.e.,

$$\boldsymbol{y = e^{3x} - 3x - 1}$$

To solve $2\dfrac{d^2y}{dx^2} - 11\dfrac{dy}{dx} + 12y = 3x - 2$, using the procedure of para 26:

Table 14.1 Form of particular integral for different functions

Type	Straightforward cases Try as particular integral:	'Snag' cases Try as particular integral
(a) $f(x) = $ a constant	$v = k$	$v = kx$ (used when CF contains a constant)
(b) $f(x) = $ polynomial (i.e. $f(x) = L + Mx + Nx^2 + \ldots$ where any of the coefficients may be zero)	$v = a + bx + cx^2 + \ldots$	
(c) $f(x) = $ an exponential function (i.e., $f(x) = Ae^{\alpha x}$)	$v = ke^{\alpha x}$	(i) $v = kxe^{\alpha x}$ (used when $e^{\alpha x}$ appears in the CF) (ii) $v = kx^2 e^{\alpha x}$ (used when $e^{\alpha x}$ and $xe^{\alpha x}$ both appear in the C.F.), etc.
(d) $f(x) = $ a sine or cosine function (i.e. $f(x) = a \sin px + b \cos px$, where a or b may be zero).	$v = A \sin px + B \cos px$	$v = x (A \sin px + B \cos px)$ (used when $\sin px$ and/or $\cos px$ appears in the (CF).
(e) $f(x) = $ a sum, e.g. (i) $f(x) = 4x^2 - 3 \sin 2x$ (ii) $f(x) = 2 - x + e^{3x}$	(i) $v = ax^2 + bx + c + d \sin 2x + e \cos 2x$ (ii) $v = ax + b + ce^{3x}$	
(f) $f(x) = $ a product, e.g. $f(x) = 2e^x \cos 2x$	$v = e^x (A \sin 2x + B \cos 2x)$	

228

(i) $2\dfrac{d^2y}{dx^2} - 11\dfrac{dy}{dx} + 12y = 3x - 2$ in D-operator form is

$(2D^2 - 11D + 12)y = 3x - 2$.

(ii) Substituting m for D gives the auxiliary equation $2m^2 - 11m + 12 = 0$

Factorising gives: $(2m - 3)(m - 4) = 0$, from which $m = \dfrac{3}{2}$ or

$m = 4$

(iii) Since the roots are real and different, the CF,
$u = Ae^{(3/2)x} + Be^{4x}$

(iv) Since $f(x) = x - 2$ is a polynomial let the PI, $v = ax + b$, (see *Table 14.1(b)*).

(v) Substituting $v = ax + b$ into $(2D^2 - 11D + 12)v = 3x - 2$ gives:

$$(2D^2 - 11D + 12)(ax + b) = 3x - 2,$$

i.e. $2D^2(ax + b) - 11D(ax + b) + 12(ax + b) = 3x - 2$
i.e. $0 - 11a + 12ax + 12b = 3x - 2$.

Equating the coefficients of x gives: $12a = 3$, from which.

$a = \dfrac{1}{4}$

Equating the constant terms gives: $-11a + 12b = -2$, i.e.

$-\dfrac{11}{4} + 12b = -2$ from which, $12b = \dfrac{3}{4}$, i.e., $b = \dfrac{1}{16}$

Hence the PI, $v = ax + b = \mathbf{\dfrac{1}{4}x + \dfrac{1}{16}}$

(vi) The general solution is given by

$$y = u + v, \text{ i.e. } \boldsymbol{y = Ae^{(3/2)x} + Be^{4x} + \dfrac{1}{4}x + \dfrac{1}{16}}.$$

To solve $\dfrac{d^2y}{dx^2} - 2\dfrac{dy}{dx} + y = 3e^{4x}$, given that when $x = 0$, $y = -\dfrac{2}{3}$ and

$\dfrac{dy}{dx} = 4\dfrac{1}{3}$, using the procedure of para 26:

(i) $\dfrac{d^2y}{dx^2} - 2\dfrac{dy}{dx} + y = 3e^{4x}$ in D-operator form is

$(D^2 - 2D + 1)y = 3e^{4x}$

(ii) Substituting m for D gives the auxiliary equation $m^2 - 2m + 1 = 0$

Factorising gives: $(m - 1)(m - 1) = 0$, from which, $m = 1$ twice.

229

(iii) Since the roots are real and equal the CF,
$u = (Ax + B)e^x$

(iv) Let the particular integral, $v = ke^{4x}$ (see *Table 14.1(c)*).

(v) Substituting $v = ke^{4x}$ into $(D^2 - 2D + 1)v = 3e^{4x}$ gives
$(D^2 - 2D + 1)ke^{4x} = 3e^{4x}$,

i.e., $D^2(ke^{4x} - 2D(ke^{4x}) + 1(ke^{4x}) = 3e^{4x}$

i.e., $16ke^{4x} - 8ke^{4x} + ke^{4x} = 3e^{4k}$.

Hence $9ke^{4k} = 3e^{4x}$, from which, $k = \dfrac{1}{3}$

Hence the PI, $v = ke^{4x} = \dfrac{1}{3}e^{4x}$

(vi) The general solution is given by $y = u + v$, i.e.

$$y = (Ax + B)e^x + \frac{1}{3}e^{4x}$$

(vii) When $x = 0$, $y = -\dfrac{2}{3}$, thus $-\dfrac{2}{3} = (0 + B)e^0 + \dfrac{1}{3}e^0$, from which, $B = -1$.

$\dfrac{dy}{dx} = (Ax + B)e^x + e^x(A) + \dfrac{4}{3}e^{4x}$

When $x = 0$, $\dfrac{dy}{dx} = 4\dfrac{1}{3}$, thus $\dfrac{13}{3} = B + A + \dfrac{4}{3}$

from which, $A = 4$, since $B = -1$.

Hence the particular solution is $y = (4x - 1)e^x + \dfrac{1}{3}e^{4x}$.

To solve $2\dfrac{d^2y}{dx^2} + 3\dfrac{dy}{dx} - 5y = 6\sin 2x$, using the procedure of para. 26:

(i) $2\dfrac{d^2y}{dx^2} + 3\dfrac{dy}{dx} - 5y = 6\sin 2x$ in D-operator form is
$(2D^2 + 3D - 5)y = 6\sin 2x$.

(ii) The auxiliary equation is $2m^2 + 3m - 5 = 0$, from which

$$(m - 1)(2m + 5) = 0, \text{ i.e., } m = 1 \text{ or } m = -\frac{5}{2}.$$

(iii) Since the roots are real and different the CF,
$u = Ae^x + Be^{-(5/2)x}$.

(iv) Let the PI, $v = C\sin 2x + D\cos 2x$ (see *Table 14.1(d)*).

(v) Substituting $v = C\sin 2x + D\cos 2x$ into
$(2D^2 + 3D - 5)v = 6\sin 2x$ gives:

$(2D^2 + 3D - 5)(C\sin 2x + D\cos 2x) = 6\sin 2x$
$D(C\sin 2x + D\cos 2x) = 2C\cos 2x - 2D\sin 2x$

$$D^2(C \sin 2x + D \cos 2x) = D(2C \cos 2x - 2D \sin 2x)$$
$$= -4C \sin 2x - 4D \cos 2x$$

Hence $(2D^2 + 3D - 5)(C \sin 2x + D \cos 2x)$
$$= -8C \sin 2x - 8D \cos 2x.$$
$$+ 6C \cos 2x - 6D \sin 2x - 5C \sin 2x - 5D \cos 2x$$
$$= 6 \sin 2x.$$

Equating coefficient of sin 2x gives: $-13C - 6D = 6$ (1)

Equating coefficients of cos 2x gives: $6C - 13D = 0$ (2)

$6 \times (1)$ gives: $-78C - 36D = 36$ (3)

$13 \times (2)$ gives: $78C - 169D = 0$ (4)

$(3) + (4)$ gives: $-205D = 36$ (5)

from which, $D = \dfrac{-36}{205}$

Substituting $D = \dfrac{-36}{205}$ into equation (1) or (2) gives

$C = \dfrac{-78}{205}$

Hence the PI, $v = \dfrac{-78}{205} \sin 2x - \dfrac{36}{205} \cos 2x$

(vi) The general solution, $y = u + v$, i.e.,

$$\boldsymbol{y = Ae^x + Be^{-5/2x} - \frac{2}{205}(39 \sin 2x + 18 \cos 2x)}$$

15 Boolean algebra and logic circuits

1   A **two-state device** is one whose basic elements can only have one or two conditions. Thus, two-way switches, which can either be on or off, and the binary numbering system, having the digits 0 and 1 only, are two-state devices. In Boolean algebra, if A represents one state, then $\bar{A}$, called 'not-A', represents the second state.

The or-function

2   In Boolean algebra, the **OR**-function for two elements A and B is written as $A + B$, and is defined as 'A, or B, or both A and B'. The equivalent electrical circuit for a two-input **OR**-function is given by two switches connected in parallel. With reference to *Figure 15.1(a)*, the lamp will be on when A is on, when B is on, or when both A and B are on. In the table shown in *Figure 15.1(b)*, all the possible switch combinations are shown in columns 1 and 2, in which a 0 represents a switch being off and a 1 represents the switch being on, these columns being called the inputs. Column 3 is called the output and a 0 represents the lamp being off and a 1 represents the lamp being on. Such a table is called a **truth table**.

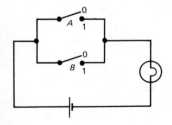

1	2	3
Input (switches)		Output (lamp)
A	B	$Z = A + B$
0	0	0
0	1	1
1	0	1
1	1	1

(a) Switching circuit for **or** − function
Figure 15.1

(b) Truth table for **or** − function

The and-function

3   In Boolean algebra, the **AND**-function for two elements A and B is written as $A.B$ and is defined as 'both A and B'. The equivalent electrical circuit for a two-input **AND**-function is given by two switches connected in series. With reference to *Figure 15.2(a)* the lamp will be on only when both A and B are on. The truth table for a two-input **AND**-function is shown in *Figure 15.2(b)*.

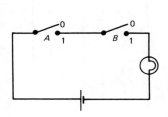

Input (switches)		Output (lamp)
A	B	$Z = A.B$
0	0	0
0	1	0
1	0	0
1	1	1

(a) Switching circuit for **and** − function   (b) Truth table for **and** − function

Figure 15.2

The not-function

4   In Boolean algebra, the **NOT**-function for element A is written as $\bar{A}$, and is defined as 'the opposite to A'. Thus if A means switch A is on. $\bar{A}$ means that switch A is off. The truth table for the **NOT**-function is shown in *Table 15.1*.

Table 15.1

Input A	Output $Z = \bar{A}$
0	1
1	0

5   In paras 2, 3 and 4 above, the Boolean expressions, equivalent switching circuits and truth tables for the three functions used in Boolean algebra are given for a two-input system. A system may have more than two inputs and the Boolean expression for a three-input **OR**-function having elements A, B and C is $A + B + C$. Similarly, a three-input **AND**-function is written as $A.B.C$. The equivalent electrical circuits and truth tables for three-input **OR** and **AND**-functions are shown in *Figures 15.3(a)* and *(b)* respectively.

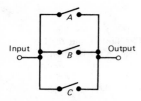

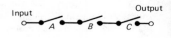

Input A B C	Output $Z = A + B + C$
0 0 0	0
0 0 1	1
0 1 0	1
0 1 1	1
1 0 0	1
1 0 1	1
1 1 0	1
1 1 1	1

(a) The **or** – function
electrical circuit and
truth table

Input A B C	Output $Z = A.B.C$
0 0 0	0
0 0 1	0
0 1 0	0
0 1 1	0
1 0 0	0
1 0 1	0
1 1 0	0
1 1 1	1

(b) The **and** – function
electrical circuit and
truth table

Figure 15.3

6   To achieve a given output, it is often necessary to use
combinations of switches connected both in series and in parallel.
If the output from a switching circuit is given by the Boolean
expression $Z = A.B + \bar{A}.\bar{B}$, the truth table is as shown in *Figure
15.4(a)*. In this table, columns 1 and 2 give all the possible
combinations of A and B. Column 3 corresponds to A.B and
column 4 to $\bar{A}.\bar{B}$, i.e., a 1 output is obtained when $A = 0$ and when
$B = 0$. Column 5 is the **OR**-function applied to columns 3 and 4
giving an output of $Z = A.B + \bar{A}.\bar{B}$. The corresponding switching
circuit is shown in *Figure 15.4(b)* in which A and B are connected
in series to give $A.B$, $\bar{A}$ and $\bar{B}$ are connected in series to give $\bar{A}.\bar{B}$,
and $A.B$ and $\bar{A}.\bar{B}$ are connected in parallel to give $A.B + \bar{A}.\bar{B}$. The
circuit symbols used are such that A means the switch is on when
A is 1, $\bar{A}$ means the switch is on when A is 0, and so on.
7   When describing a complex switching circuit by means of a
Boolean expression, often many terms and many elements per term

| 1 | 2 | 3 | 4 | 5 |
A	B	$A.B$	$\overline{A}.\overline{B}$	$Z = AB + \overline{A}.\overline{B}$
0	0	0	1	1
0	1	0	0	0
1	0	0	0	0
1	1	1	0	1

(a) Truth table for $Z = A.B + \overline{A}.\overline{B}$

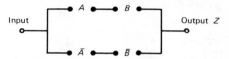

(b) Switching circuit for $Z = A.B + \overline{A}.\overline{B}$.

Figure 15.4

are used. Frequently, the laws and rules of Boolean algebra may be used to simplify the Boolean expression and some of the laws and rules are given in *Table 15.2*. These rules and laws may be verified by using truth tables.

Table 15.2

Reference	Name	Rule or Law
1	Commutative	$A+B = B+A$
2	Laws	$A.B = B.A$
3	Associative	$(A+B)+C = A+(B+C)$
4	Laws	$(A.B).C = A.(B.C)$
5	Distributive	$A.(B+C) = A.B+A.C$
6	Laws	$A+(B.C) = (A+B).(A+C)$
7		$A+0 = A$
8	Sum	$A+1 = 1$
9	Rules	$A+A = A$
10		$A+\overline{A} = 1$
11		$A.0 = 0$
12	Product	$A.1 = A$
13	Rules	$A.A = A$
14		$A.\overline{A} = 0$
15	Absorption	$A+A.B = A$
16	Rules	$A.(A+B) = A$
17		$A+\overline{A}.B = A+B$
18	Double 'not' rule	$\overline{\overline{A}} = A$

For example, $(P + \bar{P}.Q).(Q + \bar{Q}.P)$ may be simplified as follows:

Table 15.2
reference

$$(P + \bar{P}.Q).(Q + \bar{Q}.P) = P.(Q + \bar{Q}.P) + \bar{P}.Q(Q + \bar{Q}.P) \qquad 5$$

$$= P.Q + P.\bar{Q}.P + \bar{P}.Q.Q + \bar{P}.Q.\bar{Q}.P \qquad 5$$

$$= P.Q + P.\bar{Q} + \bar{P}.Q + \bar{P}.Q.\bar{Q}.P \qquad 13$$

$$= P.Q + P.\bar{Q} + \bar{P}.Q + O \qquad 14$$

$$= P.Q + P.\bar{Q} + \bar{P}.Q \qquad 7$$

$$= P.(Q + \bar{Q}) + \bar{P}.Q \qquad 5$$

$$= P.1 + \bar{P}.Q \qquad 10$$

$$= P + \bar{P}.Q \qquad 12$$

$$= P + Q \qquad 17$$

8   **De Morgan's laws** may be used to simplify **not**-functions having two or more elements. The laws state that:

(i) $\overline{A + B} = \bar{A}.\bar{B}$, and
(ii) $\overline{A.B} = \bar{A} + \bar{B}$

Each law may be verified by using a truth table. Thus, for example, to simplify the Boolean expression $(\overline{\bar{A}.B}) + (\overline{A + B})$, applying de Morgan's law (ii) to the first term gives:

$$\overline{\bar{A}.B} = \overline{\bar{A}} + \bar{B} = A + \bar{B}, \text{ since } \bar{\bar{A}} = A$$

Applying de Morgan's law (i) to the second term gives:

$$\overline{A + B} = \bar{\bar{A}}.\bar{B} = A.\bar{B}$$

Thus, $(\overline{\bar{A}.B}) + (\overline{A + B}) = (A + \bar{B}) + A.\bar{B}$
Removing the bracket and reordering gives:

$$A + A.\bar{B} + \bar{B}$$

But, by rule 15, *Table 15.2*, $A + A.B = A$.
It follows that: $A + A.\bar{B} = A$.
Thus: $(\overline{\bar{A}.B}) + (\overline{A + B}) = A + \bar{B}$.

Karnaugh maps

(i) Two-variable Karnaugh maps

9   A truth table for a two-variable expression is shown in *Table 15.3(a)*, the '1' in the third row output showing that $Z = A.\bar{B}$. Each

Table 15.3

| Inputs | | Output | Boolean |
A	B	Z	expression
0	0	0	$\overline{A}.\overline{B}$
0	1	0	$\overline{A}.B$
1	0	1	$A.\overline{B}$
1	1	0	$A.B$

(a)

A B	0 ($\overline{A}$)	1 (A)
0($\overline{B}$)	$\overline{A}.\overline{B}$	$A.\overline{B}$
1(B)	$\overline{A}.B$	$A.B$

(b)

A B	0	1
0	0	1
1	0	0

(c)

of the four possible Boolean expressions associated with a two-variable function can be depicted as shown in *Table 15.3(b)* in which one cell is allocated to each row of the truth table. A matrix similar to that shown in *Table 15.3(b)* can be used to depict $Z = A.\overline{B}$, by putting a 1 in the cell corresponding to $A.\overline{B}$ and O's in the remaining cells. This method of depicting a Boolean expression is called a two-variable Karnaugh map, and is shown in *Table 15.3(c)*.

To simplify a two-variable Boolean expression, the Boolean expression is depicted on a Karnaugh map, as outlined above. Any cells on the map having either a common vertical side or a common horizontal side are grouped together to form a **couple**. (This is a coupling together of cells, not just combining two together). The simplified Boolean expression for a couple is given by those variables common to all cells in the couple.

(ii) Three-variable Karnaugh maps

A truth table for a three-variable expression is shown in *Table 15.4(a)*, the 1's in the output column showing that:
$Z = \overline{A}.\overline{B}.C + \overline{A}.B.C + A.B.\overline{C}$.

Each of the eight possible Boolean expressions associated with a three-variable function can be depicted as shown in *Table*

Table 15.4

| | Inputs | | | Output | Boolean |
A	B	C		Z	expression
0	0	0		0	$\bar{A}.\bar{B}.\bar{C}$
0	0	1		1	$\bar{A}.\bar{B}.C$
0	1	0		0	$\bar{A}.B.\bar{C}$
0	1	1		1	$\bar{A}.B.C$
1	0	0		0	$A.\bar{B}.\bar{C}$
1	0	1		0	$A.\bar{B}.C$
1	1	0		1	$A.B.\bar{C}$
1	1	1		0	$A.B.C$

(a)

A.B C	00 $(\bar{A}.\bar{B})$	01 $(\bar{A}.B)$	11 $(A.B)$	10 $(A.\bar{B})$
$0(\bar{C})$	$\bar{A}.\bar{B}.\bar{C}$	$\bar{A}.B.\bar{C}$	$A.B.\bar{C}$	$A.\bar{B}.\bar{C}$
$1(C)$	$\bar{A}.\bar{B}.C$	$\bar{A}.B.C$	$A.B.C$	$A.\bar{B}.C$

(b)

A.B C	00	01	11	10
0	0	0	1	0
1	1	1	0	0

(c)

15.4(b), in which one cell is allocated to each row of the truth
table. A matrix similar to that shown in *Table 15.4(b)* can be used
to depict: $Z = \bar{A}.\bar{B}.C + \bar{A}.B.C + A.B.\bar{C}$, by putting 1's in the cells
corresponding to the Boolean terms on the right of the Boolean
equation and 0's in the remaining cells. This method of depicting a
three-variable Boolean expression is called a three-variable
Karnaugh map, and is shown in *Table 15.4(c)*.

To simplify a three-variable Boolean expression, the Boolean
expression is depicted on a Karnaugh map as outlined above. Any
cells on the map having common edges either vertically or
horizontally are grouped together to form couples of four cells or
two cells. During coupling the horizontal lines at the top and

238

bottom of the cells are taken as a common edge, as are the vertical lines on the left and right of the cells. The simplified Boolean expression for a couple is given by those variables common to all cells in the couple.

(iii) Four-variable Karnaugh maps

A truth table for a four-variable expression is shown in
Table 15.5(a), the 1's in the output column showing that:

$$Z = \bar{A}.\bar{B}.C.\bar{D} + \bar{A}.B.C.\bar{D} + A.\bar{B}.C.\bar{D} + A.B.C.\bar{D}$$

Each of the sixteen possible Boolean expressions associated with a four-variable function can be depicted as shown in *Table 15.5(b)*, in which one cell is allocated to each row of the truth table. A matrix similar to that shown in *Table 15.5(b)* can be used to depict

$$Z = \bar{A}.\bar{B}.C.\bar{D} + \bar{A}.B.C.\bar{D} + A.\bar{B}.C.\bar{D} + A.B.C.\bar{D}$$

by putting 1's in the cells corresponding to the Boolean terms on the right of the Boolean equation and 0's in the remaining cells. This method of depicting a four-variable expression is called a four-variable Karnaugh map, and is shown in *Table 15.5(c)*.

To simplify a four-variable Boolean expression, the Boolean expression is depicted on a Karnaugh map as outlined above. Any cells on the map having common edges either vertically or horizontally are grouped together to form couples of eight cells, four cells or two cells.

During coupling, the horizontal lines at the top and bottom of the cells may be considered to be common edges, as are the vertical lines on the left and right of the cells. The simplified Boolean expression for a couple is given by those variables common to all cells in the couple.

(iv) *Summary of procedure when simplifying a Boolean expression*
     using a Karnaugh map
  (a) Draw a four, eight or sixteen-cell matrix, depending on whether there are two, three or four variables.
  (b) Mark in the Boolean expression by putting 1's in the appropriate cells.
  (c) Form couples of 8, 4 or 2 cells having common edges forming the largest groups of cells possible. (Note that a cell containing a 1 may be used more than once when forming a couple. Also note that each cell containing a 1 must be used at least once.)
  (d) The Boolean expression for a couple is given by the variables which are common to all cells in the couple.

Table 15.5

(a)

	Inputs			Output	Boolean
A	B	C	D	Z	expression
0	0	0	0	0	$\bar{A}.\bar{B}.\bar{C}.\bar{D}$
0	0	0	1	0	$\bar{A}.\bar{B}.\bar{C}.D$
0	0	1	0	1	$\bar{A}.\bar{B}.C.\bar{D}$
0	0	1	1	0	$\bar{A}.\bar{B}.C.D$
0	1	0	0	0	$\bar{A}.B.\bar{C}.\bar{D}$
0	1	0	1	0	$\bar{A}.B.\bar{C}.D$
0	1	1	0	1	$\bar{A}.B.C.\bar{D}$
0	1	1	1	0	$\bar{A}.B.C.D$
1	0	0	0	0	$A.\bar{B}.\bar{C}.\bar{D}$
1	0	0	1	0	$A.\bar{B}.\bar{C}.D$
1	0	1	0	1	$A.\bar{B}.C.\bar{D}$
1	0	1	1	0	$A.\bar{B}.C.D$
1	1	0	0	0	$A.B.\bar{C}.\bar{D}$
1	1	0	1	0	$A.B.\bar{C}.D$
1	1	1	0	1	$A.B.C.\bar{D}$
1	1	1	1	0	$A.B.C.D$

(b)

A.B C.D	00 $(\bar{A}.\bar{B})$	01 $(\bar{A}.B)$	11 $(A.B)$	10 $(A.\bar{B})$
00 $(\bar{C}.\bar{D})$	$\bar{A}.\bar{B}.\bar{C}.\bar{D}$	$\bar{A}.B.\bar{C}.\bar{D}$	$A.B.\bar{C}.\bar{D}$	$A.\bar{B}.\bar{C}.\bar{D}$
01 $(\bar{C}.D)$	$\bar{A}.\bar{B}.\bar{C}.D$	$\bar{A}.B.\bar{C}.D$	$A.B.\bar{C}.D$	$A.\bar{B}.\bar{C}.D$
11 $(C.D)$	$\bar{A}.\bar{B}.C.D$	$\bar{A}.B.C.D$	$A.B.C.D$	$A.\bar{B}.C.D$
10 $(C.\bar{D})$	$\bar{A}.\bar{B}.C.\bar{D}$	$\bar{A}.B.C.\bar{D}$	$A.B.C.\bar{D}$	$A.\bar{B}.C.\bar{D}$

(c)

A.B C.D	0.0	0.1	1.1	1.0
0.0	0	0	0	0
0.1	0	0	0	0
1.1	0	0	0	0
1.0	1	1	1	1

Table 15.6

P Q	0	1
0	1	0
1	1	0

Table 15.7

Z \\ X.Y	0.0	0.1	1.1	1.0
0	0	1	1	0
1	1	0	0	1

For example, to simplify $\bar{P}.\bar{Q}+\bar{P}.Q$ using Karnaugh map techniques by the above procedure:

(a) the two-variable matrix is drawn and is shown in *Table 15.6*,

(b) The term $\bar{P}.\bar{Q}$ is marked with a 1 in the top left-hand cell, corresponding to $P=0$ and $Q=0$. $\bar{P}.Q$ is marked with a 1 in the bottom left-hand cell corresponding to $P=0$ and $Q=1$.

(c) The two cells containing 1's have a common horizontal edge and thus a vertical couple, shown by the broken line, can be formed.

(d) The variable common to both cells in the couple is $P=0$, i.e. $\bar{P}$, thus $\bar{P}.\bar{Q}+\bar{P}.Q=\bar{P}$.

To simplify $\bar{X}.Y.\bar{Z}+\bar{X}.\bar{Y}.Z+X.Y.\bar{Z}+X.\bar{Y}.Z$ using Karnaugh map techniques by the above procedure:

(a) a three-variable matrix is drawn and is shown in *Table 15.7*.

(b) The 1's on the matrix correspond to the expression given, i.e., for $\bar{X}.Y.\bar{Z}$, $X=0$, $Y=1$ and $Z=0$ and hence corresponds to the cell in the top row and second column, and so on.

(c) Two couples can be formed, shown by the broken lines. The couple in the bottom row may be formed since the vertical lines on the left and right of the cells are taken as a common edge.

(d) The variables common to the couple in the top row are $Y=1$ and $Z=0$, that is $Y.\bar{Z}$ and the variables common to the couple in the bottom row are $Y=0$, $Z=1$, that is $\bar{Y}.Z$. Hence

$$\bar{X}.Y.\bar{Z}+\bar{X}.\bar{Y}.Z+X.Y.\bar{Z}+X.\bar{Y}.Z=Y.\bar{Z}+\bar{Y}.Z$$

To simplify $A.B.\bar{C}.\bar{D}+A.B.C.D+\bar{A}.B.C.D+A.B.C.\bar{D}+\bar{A}.B.C.\bar{D}$ using Karnaugh map techniques, using the procedure given above, a four-variable matrix is drawn and is shown in *Table 15.8*. The 1's marked on the matrix correspond to the expression given. Two couples can be formed and are shown by the broken lines. The

241

Table 15.8

C.D \ A.B	0.0	0.1	1.1	1.0
0.0			1	
0.1				
1.1		1	1	
1.0		1	1	

four-cell couple has $B = 1$, $C = 1$, i.e. $B.C$ as the common variables to all four cells and the two-cell couple has $A.B.\bar{D}$ as the common variables to both cells. Hence, the expression simplifies to:

$$B.C + A.B.\bar{D}, \text{ i.e. } B.(C + A.\bar{D})$$

Logic circuits

10   In practice, logic gates are used to perform the **AND**, **OR** and **NOT** functions introduced in paras 2 to 4. Logic gates can be made from switches, magnetic devices or fluidic devices, but most logic gates in use are electronic devices. Various logic gates are available. For example, the Boolean expression $(A.B.C)$ can be produced using a three-input, **AND**-gate and $(C + D)$ by using a two-input **OR**-gate. The principal gates in common use are introduced in paras 11 to 15.

The term 'gate' is used in the same sense as a normal gate, the open state being indicated by a binary '1' and the closed state by a binary '0'. A gate will only open when the requirements of the gate are met and, for example, there will only be a '1' output on a two-input **AND**-gate when both the inputs to the gate are at a '1' state.

The **AND**-gate

11   Two different symbols used for a **three-input AND**-gate are shown in *Figure 15.5(a)* and the truth table is shown in *Figure 15.5(b)*. This shows that there will only be a '1' output when A is 1 and B is 1 and C is 1, written as:

$$Z = A.B.C.$$

The **OR**-gate

12   Two different symbols used for a three-input **or**-gate are shown in *Figure 15.6(a)* and the truth table is shown in *Figure*

242

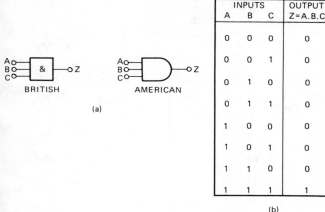

INPUTS			OUTPUT
A	B	C	Z=A.B.C
0	0	0	0
0	0	1	0
0	1	0	0
0	1	1	0
1	0	0	0
1	0	1	0
1	1	0	0
1	1	1	1

(b)

Figure 15.5

$15.6(b)$. This shows that there will be a '1' output when A is 1, or B is 1, or C is 1, or any combinations of A, B or C is 1, written as $Z = A + B + C$.

THE invert-gate or NOT-gate

13   Two different symbols used for an **invert**-gate are shown in *Figure 15.7(a)* and the truth table is shown in *Figure 15.7(b)*. This shows that a '0' input gives a '1' output and vice versa, i.e. it is an 'opposite to' function. The invert of A is written $\bar{A}$ and is called 'not-A'.

The NAND-gate

14   Two different symbols used for a **NAND**-gate are shown in *Figure 15.8(a)* and the truth table is shown in *Figure 15.8(b)*. This gate is equivalent to an **AND**-gate and an **invert**-gate in series (not-and = nand) and the output is written as $Z = \overline{A.B.C}$.

The NOR-gate

15   Two different symbols used for a **NOR**-gate are shown in *Figure 15.9(a)* and the truth table is shown in *Figure 15.9(b)*. This gate is equivalent to an **or**-gate and an **invert**-gate in series, (not-or = nor), and the output is written as:

$$Z = \overline{A + B + C}.$$

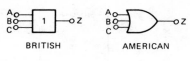

BRITISH   AMERICAN

(a)

| INPUTS | | | OUTPUT |
A	B	C	Z = A + B + C
0	0	0	0
0	0	1	1
0	1	0	1
0	1	1	1
1	0	0	1
1	0	1	1
1	1	0	1
1	1	1	1

(b)

Figure 15.6

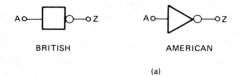

BRITISH   AMERICAN

(a)

INPUT A	OUTPUT Z = $\overline{A}$
0	1
1	0

(b)

Figure 15.7

244

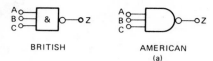

BRITISH   AMERICAN
                            (a)

| INPUTS | | | | OUTPUT |
A	B	C	A.B.C.	$Z = \overline{A.B.C.}$
0	0	0	0	1
0	0	1	0	1
0	1	0	0	1
0	1	1	0	1
1	0	0	0	1
1	0	1	0	1
1	1	0	0	1
1	1	1	1	0

(b)

Figure 15.8

Combinational logic networks

16   In most logic circuits more than one gate is needed to give
the required output. Except for the **invert**-gate, logic gates
generally have two, three or four inputs and are confined to one
function only, thus, for example, a two-input, **OR**-gate or a four-
input **AND**-gate can be used when designing a logic circuit.
For example, *Figure 15.10* shows a logic system to meet the
requirement of $Z = A.\bar{B} + C$.

With reference to *Figure 15.10* an **invert**-gate shown as (1),
gives $\bar{B}$. The **AND**-gate shown as (2), has inputs of A and $\bar{B}$, giving
$A.\bar{B}$. The **OR**-gate, shown as (3), has inputs of $A.\bar{B}$ and C, giving
$Z = A.\bar{B} + C$.

Similarly a logic system to meet the requirements of
$(P + \bar{Q}).(\bar{R} + S)$ is shown in *Figure 15.11*. The given expression shows
that two **invert**-functions are need to give $\bar{Q}$ and $\bar{R}$ and these are
shown as gates (1) and (2). Two **or**-gates, shown as (3) and (4),
give $(P + \bar{Q})$ and $(\bar{R} + S)$ respectively. Finally, an **AND**-gate, shown
as (5), gives the required output.

$$Z = (P + \bar{Q}).(\bar{R} + S).$$

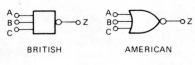

BRITISH   AMERICAN

(a)

INPUTS			A + B + C	OUTPUT $Z = \overline{A + B + C}$
A	B	C		
0	0	0	0	1
0	0	1	1	0
0	1	0	1	0
0	1	1	1	0
1	0	0	1	0
1	0	1	1	0
1	1	0	1	0
1	1	1	1	0

(b)

Figure 15.9

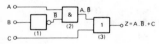

Figure 15.10

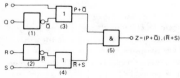

Figure 15.11

246

Universal logic gates

17   The function of any of the five logic gates in common use can be obtained by using either **NAND**-gates or **NOR**-gates and when used in this manner, the gate selected is called a **universal gate**.

(i) **NAND-gates** A single input to a **NAND**-gate gives the invert-function, as shown in *Figure 15.12(a)*.

When two **NAND**-gates are connected, as shown in *Figure 15.12(b)*, the output from the first gate is $\overline{A.B.C}$ and this is inverted by the second gate, giving $Z = \overline{\overline{A.B.C}} = A.B.C$, i.e., the **AND**-function is produced. When $\overline{A}.\overline{B}$ and $\overline{C}$ are the inputs to a **NAND**-gate, the output is $\overline{\overline{A}.\overline{B}.\overline{C}}$.

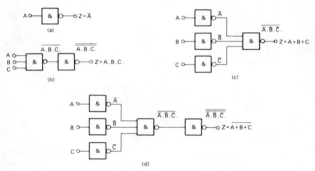

Figure 15.12

By de Morgan's law, $\overline{\overline{A}.\overline{B}.\overline{C}} = \overline{\overline{A}} + \overline{\overline{B}} + \overline{\overline{C}} = A + B + C$, i.e. a **nand**-gate is used to produce the **or**-function. The logic circuit is shown in *Figure 15.12(c)*. If the output from the logic circuit in *Figure 15.12(c)* is inverted by adding an additional **nand**-gate, the output becomes the **invert** of an **or**-function, i.e., the **nor**-function, as shown in *Figure 15.12(d)*.

When designing logic circuits, it is often easier to start at the output of the given circuit. For example, when designing a logic circuit, using **nand**-gates only having not more than three inputs, to meet the requirements of the Boolean expression $Z = \overline{A} + \overline{B} + C + \overline{D}$, the given expression shows there are four variables joined by **or**-functions.

If a four-input **NAND**-gate is used to give the required expression, the inputs are $\bar{\bar{A}}$, $\bar{\bar{B}}$, $\bar{\bar{C}}$ and $\bar{D}$, i.e. $A, B, \bar{C}$ and D. However if three inputs are not to be exceeded, two of the variables are joined and the inputs to the three-input **NAND**-gate, shown as (1) in *Figure 15.13* is $A.B$, $\bar{C}$ and D. The **AND**-function is generated by using two **NAND**-gates connected in series, shown by gates (2) and (3).

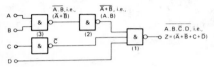

Figure 15.13

The logic circuit required to produce the given equation is as shown in *Figure 15.13*.

(ii) **NOR-gates** A single input to a **NOR**-gate gives the **invert**-function, as shown in *Figure 15.14(a)*. When two **nor**-gates are connected, as shown in *Figure 15.14(b)*, the output from the first gate is $\overline{A+B+C}$, and this is inverted by the second gate, giving $Z = \overline{\overline{A+B+C}} = A+B+C$, i.e., the **OR**-function is produced. Inputs of $\bar{A}$, $\bar{B}$ and $\bar{C}$ to a **NOR**-gate give an output of $\overline{\bar{A}+\bar{B}+\bar{C}}$.

By de Morgan's law: $\overline{\bar{A}+\bar{B}+\bar{C}} = \bar{\bar{A}}.\bar{\bar{B}}.\bar{\bar{C}} = A.B.C$, i.e., the **NOR**-gate can be used to produce the **AND**-function. The logic circuit is

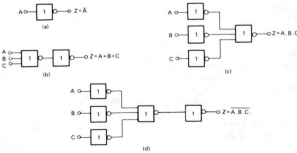

Figure 15.14

248

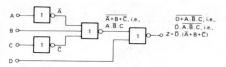

Figure 15.15

shown in *Figure 15.14(c)*. When the output of the logic circuit, shown in *Figure 15.14(c)* is inverted by adding **an additional NOR**-gate, the output then becomes the **invert** of an **OR**-function, i.e., the **NOR**-function as shown in *Figure 15.14(d)*.

For example, to design a logic circuit using nor-gates only to meet the requirements of the equation $Z = \bar{D}.(\bar{A} + B + \bar{C})$, it is usual in logic circuit design to start the design at the output. The **AND**-function between $\bar{D}$ and the terms in the bracket can be produced by using inputs of $\bar{D}$ and $\overline{\bar{A} + B + \bar{C}}$ to a **NOR**-gate, i.e. by de Morgan's law, inputs of D and $A.\bar{B}.C$. Also, inputs of $\bar{A}.B$ and $\bar{C}$ to a **NOR**-gate give an output of $\overline{\bar{A} + B + \bar{C}}$, which by de Morgan's law is $A.\bar{B}.C$. The logic circuit to produce the required equation is as shown in *Figure 15.15*.

16 Statistics

1   Data is obtained largely by two methods:

(a) by counting — for example, the number of stamps sold by a post office in equal periods of time, and

(b) by measurement — for example, the heights of a group of people.

2   When data is obtained by counting and only whole numbers are possible, the data is called **discrete**. Measured data can have any value within certain limits and is called **continuous**.

3   A **set** is a group of data and an individual value within the set is called a **member** of the set. Thus, if the masses of five people are measured correct to the nearest 0.1 kilograms and are found to be 53.1 kg, 59.4 kg, 62.1 kg, 77.8 kg and 64.4 kg, then the set of masses in kilograms for these five people is:

$$\{53.1, 59.4, 62.1, 77.8, 64.4\}$$

and one of the members of the set is 59.4.

A set containing all the members is called a **population**. Some members selected at random from a population is called a **sample**. Thus all car registration numbers form a population, but the registration numbers, of, say, 20 cars taken at random throughout the country is a sample drawn from that population.

4   The number of times that the value of a member occurs in a set is called the **frequency** of that member. Thus in the set:

$$\{2, 3, 4, 5, 4, 2, 4, 7, 9\},$$

member 4 has a frequency of three, member 2 has a frequency of 2 and the other members have a frequency of one.

5   The **relative frequency** with which any member of a set occurs is given by the ratio: $\dfrac{\text{frequency of member}}{\text{total frequency of all members}}$. For the set:

$$\{2, 3, 5, 4, 7, 5, 6, 2, 8\},$$

the relative frequency of member 5 is $\dfrac{2}{9}$. Often, relative frequency is expressed as a percentage and the **percentage relative frequency** is: (relative frequency × 100)%.

6   When the number of members in a set is comparatively small, say ten or less, the data can be represented diagrammatically without further analysis. For sets having more than ten members, those members having similar values are grouped together into **classes** to form a **frequency distribution**. To assist in accurately counting members in the various classes, a **tally diagram** is used. A frequency distribution is merely a table showing classes and their corresponding frequencies. The new set of values obtained by forming a frequency distribution is called **grouped data**.

7   The terms used in connection with grouped data are shown in *Figure 16.1(a)*. The size or range of a class is given by the **upper class boundary value** minus the **lower class boundary value**, and in *Figure 16.1* is 7.65 − 7.35, i.e., 0.3. The **class interval** for

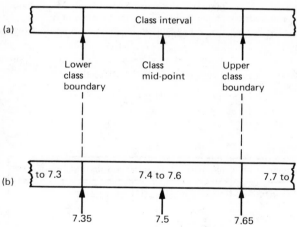

Figure 16.1

the class shown in *Figure 16.1* is 7.4 to 7.6 and the class mid-point value is given by

$$\frac{\text{upper class boundary value} - \text{lower class boundary value}}{2}$$

and in *Figure 16.1* is

$$\frac{7.65 - 7.35}{2}, \text{ i.e. } 7.5.$$

8   A cumulative frequency distribution is a table showing the cumulative frequency for each value of upper class boundary. The cumulative frequency for a particular value of upper class boundary is obtained by adding the frequency of the class to the sum of the previous frequencies.

9   Ungrouped data can be presented diagrammatically in several ways and these include:

(a) **pictograms**, in which pictorial symbols are used to represent quantities,

(b) **horizontal bar charts**, having data represented by equally spaced horizontal rectangles, and

(c) **vertical bar charts**, in which data is represented by equally spaced vertical rectangles.

For example, the number of television sets repaired in a workshop by a technician in six, one-month periods is as shown below.

Month	January	February	March	April	May	June
No. repaired	11	6	15	9	13	8

This data may be presented as a pictogram as shown in *Figure 16.2* where each symbol represents two television sets. Thus in January, $5\frac{1}{2}$ symbols are used to represent the 11 sets repaired, in February, 3 symbols are used to represent the 6 sets repaired, and so on.

The distance in miles travelled by four salesmen in a week

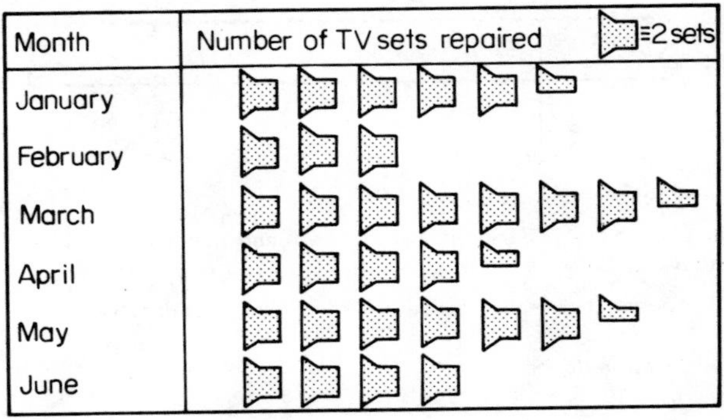

Figure 16.2

are as shown below.

Salesman	P	Q	R	S
Distance travelled (miles)	413	264	597	143

To represent this data diagrammatically by a horizontal bar chart, equally spaced horizontal rectangles of any width, but whose length is proportional to the distance travelled, are used. Thus, the length of the rectangle of salesman P is proportional to 413 miles, and so on. The horizontal bar chart depicting this data is shown in *Figure 16.3*.

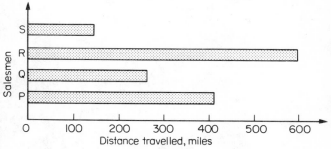

Figure 1.63

The number of issues of tools or materials from a store in a factory is observed for seven, one-hour periods in a day, and the results of the survey are as follows:

Period	1	2	3	4	5	6	7
Number of issues	34	17	9	5	27	13	6

In a vertical bar chart, equally spaced vertical rectangles of any width, but whose height is proportional to the quantity being represented, are used. Thus the height of the rectangle for period 1 is proportional to 34 units, and so on. The vertical bar chart depicting this data is shown in *Figure 16.4*.

10   Trends in ungrouped data over equal periods of time can be presented diagrammatically by a **percentage component bar chart**. In such a chart, equally spaced rectangles of any width, but whose height corresponds to 100%, are constructed. The rectangles are then sub-divided into values corresponding to the percentage relative frequencies of the members. For example, the number of various types

253

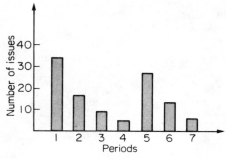

Figure 16.4

of dwellings sold by a company annually over a three-year period is as shown below.

	Year 1	Year 2	Year 3
4-roomed bungalows	24	17	7
5-roomed bungalows	38	71	118
4-roomed houses	44	50	53
5-roomed houses	64	82	147
6-roomed houses	30	30	25

To draw a percentage component bar chart to represent this data, a table of percentage relative frequency values, correct to the nearest 1%, is the first requirement. Since,

$$\text{percentage relative frequency} = \frac{\text{frequency of member} \times 100}{\text{total frequency}}$$

then for 4-roomed bungalows in year 1:

$$\text{percentage relative frequency} = \frac{24 \times 100}{24 + 38 + 44 + 64 + 30} = 12\%$$

The percentage relative frequencies of the other types of dwellings for each of the three years are similarly calculated and the results are as shown in the table below.

	Year 1	Year 2	Year 3
4-roomed bungalows	12%	7%	2%
5-roomed bungalows	19%	28%	34%
4-roomed houses	22%	20%	15%
5-roomed houses	32%	33%	42%
6-roomed houses	15%	12%	7%

254

The percentage component bar chart is produced by constructing three equally spaced rectangles of any width, corresponding to the three years. The heights of the rectangles correspond to 100% relative frequency, and are sub-divided into the values in the table of percentages shown above.

A key is used (different types of shading or different colour schemes) to indicate corresponding percentage values in the rows of the table of percentages. The percentage component bar chart is shown in *Figure 16.5*.

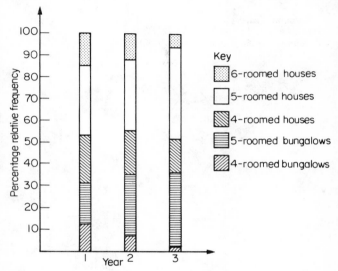

Figure 16.5

11   A **pie diagram** is used to show diagrammatically the parts making up the whole. In a pie diagram, the area of a circle represents the whole, and the areas of the sectors of the circle are made proportional to the parts which make up the whole.

For example, the retail price of a product costing £2 is made up as follows: materials 10p, labour 20p, research and development 40p, overheads 70p, profit 60p. To present this data on a pie diagram, a circle of any radius is drawn, and the area of the circle represents the whole, which in this case is £2. The circle is sub-divided into sectors so that the areas of the sectors are proportional to the parts, i.e. the parts which make up the total retail price. For the area of a sector to be proportional to a part, the angle at the centre of the circle must be proportional to that part. The whole, £2 or 200p, corresponds to 360°. Therefore,

$$10\text{p corresponds to } 360 \times \frac{10}{200} \text{ degrees, i.e. } 18°$$

$$20\text{p corresponds to } 360 \times \frac{20}{200} \text{ degrees, i.e. } 36°$$

and so on, giving the angles at the centre of the circle for the parts of the retail prices as: 18°, 36°, 72°, 126° and 108° respectively. The pie diagram is shown in *Figure 16.6*.

1p ≡ 1.8°

Figure 16.6

12 Grouped data may be presented diagrammatically in three ways:
    (a) by using a **frequency polygon**, which is the graph product by plotting frequency against class mid-point values and joining the co-ordinates with straight lines,
    (b) by producing a **histogram**, in which the areas of vertical adjacent rectangles are made proportional to the frequencies of the classes, and
    (c) by drawing an **ogive** or **cumulative frequency distribution curve**, which is a graph produced by plotting cumulative frequency against upper class boundary values and joining the co-ordinates by straight lines.

For example, the masses of 50 ingots in kilograms are measured correct to the nearest 0.1 kg and the results are as shown below:

```
8.0  8.6  8.2  7.5  8.0  9.1  8.5  7.6  8.2  7.8
8.3  7.1  8.1  8.3  8.7  7.8  8.7  8.5  8.4  8.5
7.7  8.4  7.9  8.8  7.2  8.1  7.8  8.2  7.7  7.5
8.1  7.4  8.8  8.0  8.4  8.5  8.1  7.3  9.0  8.6
7.4  8.2  8.4  7.7  8.3  8.2  7.9  8.5  7.9  8.0
```

The range of the data is the member having the largest value minus the member having the smallest value. Inspection of the set of data shows that:

$$\text{range} = 9.1 - 7.1 = 2.0$$

The size of each class is given approximately by

$$\frac{\text{range}}{\text{number of classes}}$$

If about 7 classes are required, the size of each class is 2.0/7, that is, approximately 0.3

To assist with accurately determining the number in each class, a tally diagram is produced as shown in *Table 16.1(a)*. This is obtained by listing the classes in the left-hand column and then inspecting each of the 50 members of the set of data in turn and allocating it to the appropriate class by putting a '1' in the appropriate row. Each fifth '1' allocated to a particular row is marked as an oblique line to help with final counting.

A **frequency distribution** for the data is shown in *Table 16.1(b)* and lists classes and their corresponding frequencies. Class

Table 16.1

Class	Tally
7.1 to 7.3	III
7.4 to 7.6	⊞
7.7 to 7.9	⊞ IIII
8.0 to 8.2	⊞ ⊞ IIII
8.3 to 8.5	⊞ ⊞ I
8.6 to 8.8	⊞ I
8.9 to 9.1	II

(a) Tally diagram

Class	Class mid-point	Frequency
7.1 to 7.3	7.2	3
7.4 to 7.6	7.5	5
7.7 to 7.9	7.8	9
8.0 to 8.2	8.1	14
8.3 to 8.5	8.4	11
8.6 to 8.8	8.7	6
8.9 to 9.1	9.0	2

(b) Frequency distribution

257

mid-points are also shown in this table, since they are used when constructing a frequency polygon and histogram.

A **frequency polygon** is shown in *Figure 16.7(a)*, the co-ordinates corresponding to the class mid-point/frequency values, given in *Table 16.1(b)*. The co-ordinates are joined by straight lines and the polygon is 'anchored-down' at each end by joining to the next class mid-point value and zero frequency.

A **histogram** is shown in *Figure 16.7(b)*, the width of a rectangle corresponding to (upper class boundary value — lower class boundary value) and height corresponding to the class

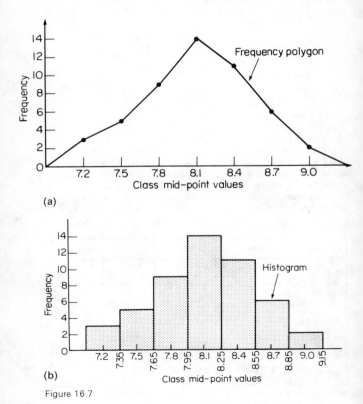

Figure 16.7

258

frequency. The easiest way to draw a histogram is to mark class mid-point values on the horizontal scale and to draw the rectangles symmetrically about the appropriate class-midpoint values and touching one another. A histogram for the data given in *Table 16.1(d)* is shown in *Figure 16.7(b)*.

A cumulative frequency distribution is a table giving values of cumulative frequency for the values of upper class boundaries, and is shown in *Table 16.2*. Columns 1 and 2 show the classes and their frequencies. Column 3 lists the upper class boundary values for the classes given in column 1. Column 4 gives the cumulative frequency values for all frequencies less than the upper class boundary values given in column 3. Thus, for example, for the 7.7 to 7.9 class shown in row 3, the cumulative frequency value is the sum of all frequencies having values of less than 7.95, i.e., $3 + 5 + 9 = 17$, and so on.

Table 16.2

1 Class	2 Frequency	3 Upper class boundary	4 Cumulative frequency
		Less than	
7.1–7.3	3	7.35	3
7.4–7.6	5	7.65	8
7.7–7.9	9	7.95	17
8.0–8.2	14	8.25	31
8.3–8.5	11	8.55	42
8.6–8.8	6	8.85	48
8.9–9.1	2	9.15	50

The ogive for the cumulative frequency distribution given in *Table 16.2* is shown in *Figure 16.8*. The co-ordinates corresponding to each upper class boundary/cumulative frequency value are plotted and the co-ordinates are joined by straight lines. (*Note*: not the best curve drawn through the co-ordinates as in experimental work.) The ogive is 'anchored' at its start by adding the co-ordinate $(7.05, 0)$.

Measures of central tendency and dispersion

Measures of central tendency

13   A single value, which is representative of a set of values, may be used to given an indication of the general size of the members in a set, the word 'average' often being used to indicate the single

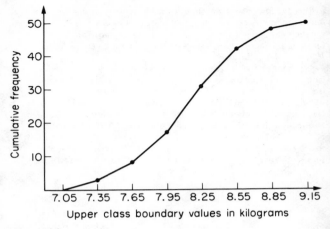

Figure 16.8

value. The statistical term used for 'average' is the arithmetic mean or just the mean. Other measures of central tendency may be used and these include the median and the modal values.

Discrete data

14   The **arithmetic mean value** is found by adding together the values of the members of a set and dividing by the number of members in the set. Thus, the mean of the set of numbers:

$\{4, 5, 6, 9\}$

is:

$\dfrac{4 + 5 + 6 + 9}{4}$, i.e. 6

In general, the mean of the set:

$\{x_1, x_2, x_3, \ldots, x_n\}$

is

$$\bar{x} = \frac{x_1 + x_2 + x_3 + \ldots x_n}{n} \text{ , written as } \frac{\sum x}{n},$$

260

where $\sum$ is the Greek letter 'sigma' and means 'the sum of', and $\bar{x}$, (called x-bar), is used to signify a mean value.

15 The **median value** often gives a better indication of the general size of a set containing extreme values. The set: $\{7, 5, 74, 10\}$ has a mean value of 24, which is not really representative of any of the values of the members of the set. The median value is obtained by:

(a) **ranking** the set in ascending order of magnitude, and
(b) selecting the value of the middle member for sets containing an odd numbers of members, or finding the value of the mean of the two middle members for sets containing an even number of members.

For example, the set $\{7, 5, 74, 10\}$ is ranked as $\{5, 7, 10, 74\}$, and since it contains an even number of members (four in this case), the mean of 7 and 10 is taken, giving a median value of 8.5. Similarly, the set: $\{3, 81, 15, 7, 14\}$ is ranked as $\{3, 7, 14, 15, 81\}$ and the median value is the value of the middle number, i.e. 14.

16 The **modal value**, or just the **mode**, is the most commonly occurring value in a set. If two values occur with the same frequency, the set is 'bi-modal'. The set

$$\{5, 6, 8, 2, 5, 4, 6, 5, 3\}$$

has a modal value of 5, since the member having a value of 5 occurs three times.

Grouped data

17 The mean value for a set of grouped data is found by determining the sum of the (frequency × class mid-point values) and dividing by the sum of the frequencies, i.e. mean value

$$\bar{x} = \frac{f_1 x_1 + f_2 x_2 + \ldots + f_n x_n}{f_1 + f_2 + \ldots + f_n} = \frac{\sum (fx)}{\sum f},$$

where f is the frequency of the class having a mid-point value of x, and so on.

For example, the frequency distribution for the value of resistance in ohms of 48 resistors is:

20.5 − 20.9	3,	21.0 − 21.4	10,	21.5 − 21.9	11
22.0 − 22.4	13,	22.5 − 22.9	9,	23.0 − 23.4	2

The class mid-point/frequency values are:

20.7	3,	21.2	10,	21.7	11,	22.2	13,	22.7	9
and		23.2	2.						

For grouped data, the mean value is given by:

$$\bar{x} = \frac{\sum(fx)}{\sum f}$$

where f is the class frequency and x is the class mid-point value. Hence, mean value,

$$\bar{x} = \frac{(3 \times 20.7) + (10 \times 21.2) + (11 \times 21.7) + (13 \times 22.2) + (9 \times 22.7) + (2 \times 23.2)}{48}$$

$$= \frac{1052.1}{48} = 21.929\ldots$$

i.e. **the mean value is 21.9 ohms**, correct to 3 significant figures.
18 The mean, median and modal values for grouped data may be determined from a histogram. In a histogram, frequency values are represented vertically and variable values horizontally.

The mean value is given by the value of the variable corresponding to a vertical line drawn through the centroid of the histogram.

The median value is obtained by selecting a variable value such that the area of the histogram to the left of a vertical line drawn through the selected variable value is equal to the area of the histogram on the right of the line.

The modal value is the variable value obtained by dividing the width of the highest rectangle in the histogram in proportion to the heights of the adjacent rectangles.

For example, the time taken in minutes to assemble a device is measured 50 times and the results are as shown below:

14.5 − 15.5	5,	16.5 − 17.5	8,	18.5 − 19.5	16
20.5 − 21.5	12,	22.5 − 23.5	6,	24.5 − 25.5	3

The mean, median and modal values of this distribution may be determined from a histogram depicting the data.

The histogram is shown in *Figure 16.9*. The mean value lies at the centroid of the histogram. With reference to any arbitrary axis, say YY shown at a time of 14 minutes, the position of the horizontal value of the centroid can be obtained from the relationship $AM = \sum(am)$, where A is the area of the histogram, M is the horizontal distance of the centroid from the axis YY, a is the area of a rectangle of the histogram and m is the distance of the centroid of the rectangle from YY. The areas of the individual rectangles are shown circled on the histogram giving a total area of 100 square units. The positions, m, of the centroids of the indvidual rectangles are 1, 3, 5, ... units from YY.

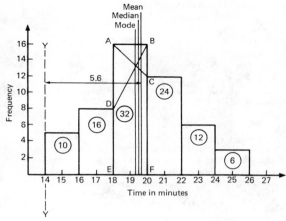

Figure 16.9

Thus $100M = (10 \times 1) + (16 \times 3) + (32 \times 5) + (24 \times 7) + (12 \times 9)$
$+ (6 \times 11)$ i.e. $M = \dfrac{560}{100} = 5.6$ units from YY.

Thus the position of the mean with reference to the time scale is
$14 + 5.6$, i.e. **19.6 minutes**.

The median is the value of time corresponding to a vertical
line dividing the total area of the histogram into two equal parts.
The total area is 100 square units, hence the vertical line must be
drawn to give 50 units of area on each side. To achieve this with
reference to *Figure 16.9*, rectangle ABFE must be split so that
$50 - (10 + 16)$ units of area lie on one side and $50 - (24 + 12 + 6)$
units of area lie on the other. This shows that the area of ABFE is
split so that 24 units of area lie to the left of the line and 8 units of
area lie to the right, i.e. the vertical line must pass through 19.5
minutes. Thus the median value of the distribution is **19.5 minutes**.

The mode is obtained by dividing the line AB, which is the
height of the highest rectangle, proportionally to the heights of the
adjacent rectangles. With reference to *Figure 16.9*, this is done by
joining AC and BD and drawing a vertical line through the point
of intersection of these two lines. This gives the mode of the
distribution and is **19.3 minutes**.

19   The **standard deviation** of a set of data gives an indication of the amount of dispersion, or the scatter, of members of the set from the measure of central tendency. Its value is the root-mean-square value of the members of the set and for discrete data is obtained as follows:

(a) determine the measure of central tendency, usually the mean value, $\bar{x}$, (occasionally the median or modal value are specified),

(b) calculate the deviation of each member of the set from the mean, giving $(x_1 - \bar{x})$, $(x_2 - \bar{x})$, $(x_3 - \bar{x})$, ...

(c) determine the squares of these deviations, i.e., $(x_1 - \bar{x})^2$, $(x_2 - \bar{x})^2$, $(x_3 - \bar{x})^2$, ...

(d) find the sum of the squares of the deviations, that is $(x_1 - \bar{x})^2 + (x_2 - \bar{x})^2 + (x_3 - \bar{x})^2 + \ldots$,

(e) divide by the number of members in the set, n, giving

$$\frac{(x_1 + \bar{x})^2 + (x_2 + \bar{x})^2 + (x_3 + \bar{x})^2 + \ldots}{n}$$

(f) determine the square root of (e).

The standard deviation is indicated by σ (the Greek letter small 'sigma') and is written mathematically as:

$$\text{standard deviation } \sigma = \sqrt{\left\{ \frac{\sum (x - \bar{x})^2}{n} \right\}},$$

where x is a member of the set, $\bar{x}$ is the mean value of the set and n is the number of members in the set. The value of standard deviation gives an indication of the distance of the members of a set from the mean value. The set:

$$\{1, 4, 7, 10, 13\}$$

has a mean value of 7 and a standard deviation of about 4.2. The set

$$\{5, 6, 7, 8, 9\}$$

also has a mean value of 7, but the standard deviation is about 1.4. This shows that the members of the second set are mainly much closer to the mean value than the members of the first set.

For example, to determine the standard deviation from the mean of the set of numbers:

$$\{5, 6, 8, 4, 10, 3\},$$

correct to 4 significant figures:

The arithmetic mean, $\bar{x} = \dfrac{\sum x}{n} = \dfrac{5+6+8+4+10+3}{6} = 6$

Standard deviation, $\sigma = \sqrt{\left\{\dfrac{\sum(x-\bar{x})^2}{n}\right\}}$

The $(x-\bar{x})^2$ values are: $(5-6)^2$, $(6-6)^2$, $(8-6)^2$, $(4-6)^2$, $(10-6)^2$ and $(3-6)^2$. The sum of the $(x-\bar{x})^2$ values, i.e., $\sum(x-\bar{x})^2 = 1+0+4+4+16+9$ i.e.,

$$\sum(x-\bar{x})^2 = 34.$$

and

$$\dfrac{\sum(x-\bar{x})^2}{n} = \dfrac{34}{6} = 5.6, \text{ since there are 6 members in the set.}$$

Hence, standard deviation,

$$\sigma = \sqrt{\left\{\dfrac{\sum(x-\bar{x})^2}{n}\right\}}$$

$$= \sqrt{5.6} = \mathbf{2.380}, \text{ correct to 4 significant figures.}$$

20 For grouped data, standard deviation,

$$\sigma = \sqrt{\left[\dfrac{\sum\{f(x-\bar{x})^2\}}{\sum f}\right]}$$

where f is the class frequency value, x is the class mid-point value and $\bar{x}$ is the mean value of the grouped data. For example, the frequency distribution for the values of resistance in ohms of 48 resistors is:

20.5 − 20.9	3,	21.0 − 21.4	10,	21.5 − 21.9	11,
22.0 − 22.4	13,	22.5 − 22.9	9,	23.0 − 23.4	2.

From para 17, the distribution mean value, $\bar{x} = 21.92$, correct to 4 significant figures. The 'x-values' are the class mid-point values, i.e. 20.7, 21.2, 21.7, ... Thus the $(x-\bar{x})^2$ values are $(20.7-21.92)^2$, $(21.2-21.92)^2$ $(21.7-21.92)^2$, ..., and the $f(x-\bar{x})^2$ values are $3(20.7-21.92)^2$, $10(21.2-21.92)^2$, ... The $\sum f(x-\bar{x})^2$ values are $4.4652 + 5.1840 + 0.5324 + 1.0192 + 5.4756 + 3.2768$, i.e. 19.9532.

$$\dfrac{\sum\{f(x-\bar{x})^2\}}{\sum f} = \dfrac{19.9532}{48} = 0.415\ 69$$

and the standard deviation,

$$\sigma = \sqrt{\left\{\frac{\sum\{f(x-\bar{x})^2\}}{\sum f}\right\}}$$

$$= \sqrt{0.415\,69} = 0.645, \text{ correct to 3 significant figures.}$$

21   Other measures of dispersion which are sometimes used are the quartile, decile and percentile values. The **quartile values** of a set of discrete data are obtained by selecting the values of members which divide the set into four equal parts. Thus for the set:

$$\{2, 3, 4, 5, 5, 7, 9, 11, 13, 14, 17\}$$

there are 11 members and the values of the members dividing the set into four equal parts are 4, 7 and 13. These values are signified by Q_1, Q_2 and Q_3 and called the first, second and third quartile values respectively. It can be seen that the second quartile value, Q_2, is the value of the middle member and hence is the median value of the set.

22   For grouped data the ogive may be used to determine the quartile values. In this case, points are selected on the vertical cumulative frequency values of the ogive, such that they divide the total value of cumulative frequency into four equal parts. Horizontal lines are drawn from these values to cut the ogive. The values of the variable corresponding to these cutting points on the ogive give the quartile values.

For example, the frequency distribution given below refers to the overtime worked by a group of craftsmen during each of 48 working weeks in a year.

25–29	5,	30 – 34	4,	35 – 39	7,	40 – 44	11
45–49	12,	50 – 54	8,	55 – 59	1.		

The cumulative frequency distribution (i.e. upper class boundary/cumulative frequency values) is:

29.5	5,	34.5	9,	39.5	16,	44.5	27,	49.5	39,
54.5	47,	59.5	48						

The ogive is formed by plotting these values on a graph, as shown in *Figure 16.10*. The total frequency is divided ito four equal parts, each having a range of 48/4, i.e. twelve. This gives cumulative frequency values of 0 to 12 corresponding to the first quartile 12 to 24 corresponding to the second quartile 24 to 36 corresponding to the third quartile and 36 to 48 corresponding to the fourth quartile of the distribution, i.e. the distribution is divided into four equal parts.

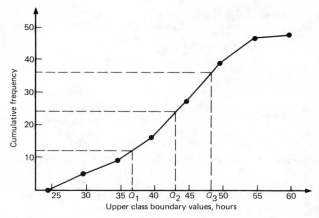

Figure 16.10

The quartile values are those of the variable corresponding to cumulative frequency values of 12, 24 and 36, marked Q_1, Q_2 and Q_3 in *Figure 16.10*. These values, correct to the nearest hour, are **37 hours**, **43 hours** and **48 hours** respectively. The Q_2 value is also equal to the median value of the distribution. One measure of the dispersion of a distribution is called the **semi-interquartile range** and is given by $(Q_3 - Q_1)/2$, and is $(48 - 37)/2$ in this case,

i.e., $5\frac{1}{2}$ **hours**.

23 When a set contains a large number of members, the set can be split into ten parts, each containing an equal number of members. These ten parts are then called **deciles**. For sets containing a very large number of members, the set may be split into one hundred parts, each containing an equal number of members. One of these parts is called a **percentile**.

Probability

24 The probability of something happening is the likelihood or chance of it happening. Values of probability lie between 0 and 1, where 0 represents an absolute impossibility and 1 represents an absolute certainty. The probability of an event happening usually lies somewhere between these two extreme values and is expressed either as a proper or a decimal fraction.

Examples of probability are:

that a length of copper wire has zero resistance at $100°C$ — 0

that a fair, six-sided dice will stop with a 3 upwards — $\frac{1}{6}$ or 0.1667

that a fair coin will land with a head upwards — $\frac{1}{2}$ or 0.5

that a length of copper wire has some resistance at $100°C$ — 1

25 If p is the probability of an event happening and q is the probability of the same event not happening, then the total probability is $p + q$ and is equal to unity, since it is an absolute certainty that the event either does or does not occur, i.e. $p + q = 1$.

26 The **expectation**, E, of an event happening is defined in general terms as the product of the probability p of an event happening and the number of attempts made, n, i.e., $E = pn$. Thus, since the probability of obtaining a 3 upwards when rolling a fair dice is $\frac{1}{6}$, the expectation of getting a 3 upwards on four throws of the dice is $\frac{1}{6} \times 4$, i.e., $\frac{2}{3}$. Thus expectation is the average occurrence of an event.

27 A **dependent event** is one in which the probability of one event happening affects the probability of another event happening. Let 5 transistors be taken at random from a batch of 100 transistors for test purposes, and the probability of there being a defective transistor, p_1, be determined. At some later time, let another 5 transistors be taken at random from the 95 remaining transistors in the batch and the probability of there being a defective transistor, p_2, can be determined. The value of p_2 is different from p_1 since the batch size has effectively altered from 100 to 95, i.e., the probability p_2 is dependent on probability p_1. Since 5 transistors are drawn, and then another 5 transistors are drawn without replacing the first 5, the second random selection is said to be **without replacement**.

28 An **independent event** is one in which the probability of an event happening does not affect the probability of another event happening. If 5 transistors are taken at random from a batch of transistors and the probability of a defective transistor p_1 is determined and the process is repeated after the original 5 have been replaced in the batch to give p_2, then p_1 is equal to p_2. Since the 5 transistors are replaced between draws, the second selection is said to be **with replacement**.

The addition law of probability

29   The addition law or probability is recognised by the word
'**or**' joining the probabilities. If p_A is the probability of event A
happening and p_B is the probability of event B happening, the
probability of event A **or** event B happening is given by $p_A + p_B$.
Similarly, the probability of events A **or** B **or** C **or** ... N
happening is given by: $p_A + p_B + p_C + \ldots p_N$.

The multiplication law of probability

30   The multiplication law of probability is recognised by the
word '**and**' joining the probabilities. If p_A is the probability of
event A happening and p_B is the probability of event B happening,
the probability of event A **and** event B happening is given by
$p_A \times p_B$. Similarly, the probability of events A **and** B **and** C **and** ...
N happening is given by $p_A \times p_B \times p_C \times \ldots p_N$.

31   The addition and multiplication laws of probability may be
combined as shown below. Let p_A, p_B and p_C be the probabilities of
events A, B and C respectively happening. The probabilities of
events A, B and C **not** happening may be shown as $\bar{p}_A$, $\bar{p}_B$ and $\bar{p}_C$,
(where $p_A + \bar{p}_A = 1$, see para 25). Then for example:

(i) the probability of events (A **and** B) **or** C happening is
$(p_A \times p_B) + p_C$,
(ii) the probability of (event A **or** event B happening)
and event C happening is $(p_A + p_B) \times p_C$,
(iii) the probability of (events A **and** B happening) **or**
(event A happening and event C not happening) is
$(p_A \times p_B) + (p_A + \bar{p}_C)$,
(iv) the probability of (event A **and** B not happening) **or**
(event C happening) is $(\bar{p}_A \times \bar{p}_B) + p_C$, and so on.

For example, to determine the probability of selecting at random
the winning horse in a race in which 10 horses are running:

Since only one of the ten horses can win, the probability of
selecting at random the winning horse is:

$$\frac{\text{number of winners}}{\text{number of horses}}, \text{ i.e., } \frac{1}{10}$$

To determine the probability of selecting at random the winning
horse in the first race or the winning horse in the second race if
there are 10 horses in each race:

The probability of selecting at random the winning horse in the
first race or the winning horse in the second race is given by the
addition law of probability, since the word or joins the two

probabilities. From above, the probability of selecting the winning horse is 1/10 for the first race and also 1/10 for the second race. Hence the probability of selecting the winning horse from the first or the second race is

$$\frac{1}{10} + \frac{1}{10} = \mathbf{\frac{1}{5}}$$

To determine the probability of selecting at random the winning horses in both the first and second races if there are 10 horses in each race:

The probability of selecting the winning horse in the first race is $\frac{1}{10}$.

The probability of selecting the winning horse in the second rate is $\frac{1}{10}$.

The probability of selecting the winning horse in the first **and** second race is given by the multiplication law or probability,

i.e. probability $= \frac{1}{10} \times \frac{1}{10} = \mathbf{\frac{1}{100}}$ **or 0.01**

Let a batch of 40 components contain 5 which are defective. If a component is drawn at random from the batch and tested and then a second component is drawn at random, the probability of having one defective component, both with and without replacement is determined as follows:

The probability of having one defective component can be achieved in two ways. If p is the probability of drawing a defective component and q is the probability of drawing a satisfactory component, then the probability of having one defective component is given by drawing a satisfactory component and then a defective component or by drawing a defective component and then a satisfactory one, i.e., by $q \times p + p \times q$.

With replacement:

$$p = \frac{5}{40} = \frac{1}{8} \text{ and } q = \frac{35}{40} = \frac{7}{8}$$

Hence, the probability of having a defective component is

$$\frac{1}{8} \times \frac{7}{8} + \frac{7}{8} \times \frac{1}{8}, \text{ i.e., } \frac{7}{64} + \frac{7}{64} = \frac{14}{64} = \mathbf{\frac{7}{32}} \text{ or } \mathbf{0.2188}.$$

Without replacement:

$p_1 = \dfrac{1}{8}$ and $q_1 = \dfrac{7}{8}$ on the first of the two draws. The batch number

is now 39 for the second draw, thus $p_2 = \dfrac{5}{39}$ and $q_2 = \dfrac{35}{39}$.

$$p_1 q_2 + q_1 p_2 = \frac{1}{8} \times \frac{35}{39} + \frac{7}{8} \times \frac{5}{39} = \frac{35+35}{312} = \mathbf{\frac{70}{312}} \text{ or } \mathbf{0.2244}$$

The Binomial distribution

32   The binomial distribution deals with two numbers only, these being the probability that an event will happen, p, and the probability that an event will not happen, q. Thus, when a coin is tossed, if p is the probability of the coin landing with a head upwards, q is the probability of the coin landing with a tail upwards. $p + q$ must always be equal to unity. A binomial distribution can be used for finding, say, the probability of getting three heads in seven tosses of the coin, or in industry for determining defect rates as a result of sampling.

33   One way of defining a binomial distribution is as follows:

if p is the probability that an event will happen and q is the probability that the event will not happen, then the probabilities that the event will happen 0, 1, 2, 3, ..., n times in n trials are given by the successive terms of the expansion of $(q + p)^n$, taken from left to right

The binomial expansion is used to obtain the terms of $(q + p)^n$. For example, to determine the probability of having at least 1 girl in a family of 4 children, assuming equal probability of male and female birth:

The probability of a girl being born, p is 0.5 and the probability of a girl not being born (male birth), q, is also 0.5. The number in the family, n, is 4. From above, the probabilities of 0, 1, 2, 3, 4 girls in a family of 4 are given by the successive terms of the expansion of $(q + p)^4$ taken from left to right.

From section 4, page 41, $(q + p)^4 = q^4 + 4q^3 p + 6q^2 p^2 + 4qp^3 + p^4$.
Hence the probability of no girls is q^4, i.e., 0.5^4 $= 0.0625$

    the probability of 1 girl is $4q^3 p$, i.e., $4 \times 0.5^3 \times 0.5$ $= 0.2500$

    the probability of 2 girls is $6q^2 p^2$, i.e., $6 \times 0.5^2 \times 0.5^2 = 0.3750$

    the probability of 3 girls is $4qp^3$, i.e., $4 \times 0.5 \times 0.5^3$ $= 0.2500$

    the probability of 4 girls is p^4, i.e., 0.5^4 $= 0.0625$

       Total probability, $(q + p)$ $= 1.0000$

The probability of having at least one girl is the sum of the probabilities of having 1, 2, 3 and 4 girls, i.e.,

$$0.2500 + 0.3750 + 0.2500 + 0.0625 = 0.9375$$

[Alternating, the probability of having at least 1 girl is:
$1 - $ (the probability of having no girls), i.e., $1 - 0.0625$, giving 0.9375, as obtained previously.]
Similarly, the probability of having at least 1 girl and 1 boy in a family of 4 is given by the sum of the probabilities of having: 1 girl and 3 boys, 2 girls and 2 boys and 3 girls and 1 boy, i.e.,

$$0.2500 + 0.3750 + 0.2500 = 0.8750$$

[Alternatively, this is also the probability of having $1 - $ (probability of having no girls + probability of having no boys), i.e., $1 - 2 \times 0.0625 = 0.8750$, as obtained previously.]

34   In industrial inspection, p is often taken as the probability that a component is defective and q is the probability that the component is satisfactory. In this case, a binomial distribution may be defined as:

the probabilities that 0, 1, 2, 3, ..., n components are defective in a sample of n components, drawn at random from a large batch of components, are given by the successive terms of the expansion of $(q + p)^n$, taken from left to right.

For example, a package contains 50 similar components and inspection shows that four have been damaged during transit. Let six components be drawn at random from the contents of the package.

The probability of a component being damaged, p, is 4 in 50, i.e. 0.08 per unit. Thus the probability of a component not being damaged, q is $1 - 0.08$, i.e. 0.92. From above, the probability of there being 0, 1, 2, ..., 6, damaged components is given by the successive terms of $(q + p)^6$, taken from left to right.

$$(q + p)^6 = q^6 + 6q^5p + 15q^4p^2 + 20q^3p^3 + \ldots$$

Thus, for example, the probability of one damaged component is

$$6q^5p = 6 \times 0.92^5 \times 0.08, \text{ i.e., } 0.3164,$$

and the probability of less than three damaged components is given by the sum of the probabilities of 0, 1 and 2 damaged components.

$$q^6 + 6q^5p + 15q^4p^2 = 0.92^6 + 6 \times 0.92^5 \times 0.08 + 15 \times 0.92^4 \times 0.08^2$$

$$= 0.6064 + 0.3164 + 0.0688 = \mathbf{0.9916}$$

272

The Poisson distribution

(i) When the number of trials, n, in a binomial distribution becomes large, (usually taken as larger than 10), the calculations associated with determining the values of the terms becomes laborious. If n is large, p is small and the product np is less than 5, a very good approximation to a binomial distribution is given by the corresponding poisson distribution, in which calculations are usually simpler.

(ii) The Poisson approximation to a binomial distribution may be defined as follows:

'the probabilities that an event will happen 0, 1, 2, 3, ..., n times in n trials are given by the successive terms of the expression

$$e^{-\lambda}\left(1 + \lambda + \frac{\lambda^2}{2!} + \frac{\lambda^3}{3!} + \dots\right) \text{ taken from left to right'.}$$

The symbol λ is the expectation of an event happening and is equal to np.

For example, let 3% of the gearwheels produced by a company be defective and let a sample of 80 gearwheels be taken.

The sample number, n, is large, the probability of a defective gearwheel, p, is small and the product np is 80×0.03, i.e., 2.4, which is less than 5. Hence a Poisson approximation to a binomial distribution may be used. The expectation of a defective gearwheel, $\lambda = np = 2.4$.

From above, the probabilities of 0, 1, 2, ... defective gearwheels are given by the successive terms of the expression

$$e^{-\lambda}\left(1 + \lambda + \frac{\lambda^2}{2!} + \dots\right), \text{ taken from left to right,}$$

i.e., by $e^{-\lambda}$, $\lambda e^{-\lambda}$, $\dfrac{\lambda^2 e^{-\lambda}}{2!} \dots$ Thus:

probability of no defective gearwheels is

$e^{-\lambda} = e^{-2.4} = 0.0907$

probability of 1 defective gearwheel is

$\lambda e^{-\lambda} = 2.4 e^{-2.4} = 0.2177$

probability of 2 defective gearwheels is

$$\frac{\lambda^2 e^{-\lambda}}{2!} = \frac{2.4^2 e^{-2.4}}{2 \times 1} = 0.2613$$

Thus, the porbability of having say 2 defective gearwheels is **0.2613**, and the probability of having more than 2 defective gearwheels is 1 — (the sum of the probabilities of having 0, 1 and 2 defective gearwheels), i.e.,
1 — (0.0907 + 0.2177 + 0.2613), that is, **0.4303**.

36   The principal use of a Poisson distribution is to determine the theoretical probabilities when p, the probability of an event happening, is known, but q, the probability of the event not happening is not known. For example, the average number of goals scored per match by a football team can be calculated, but it is not possible to quantify the number of goals which were not scored. In this type of problem, a Poisson distribution may be defined as follows:

'the probabilities of an event occurring 0, 1, 2, 3, ... times are given by the successive terms of the expression $e^{-\lambda}\left(1 + \lambda + \dfrac{\lambda^2}{2!} + \dfrac{\lambda^3}{3!} + \dots\right)$. *taken from left to right'*.

The symbol λ is the value of the average occurrence of the event. For example, a production department has 35 similar milling machines. The number of breakdowns on each machine averages 0.06 per week. Since the average occurrence of a breakdown is known but the number of times when a machine did not break down is not known, a Poisson distribution must be used to determine, say, the probability of having one machine breaking down in a week.

The expectation of a breakdown for 35 machines is 35×0.06, i.e., 2.1 breakdowns per week. From above, the probabilities of a breakdown occurring 0, 1, 2, ... times are given by the successive terms of the expression $e^{-\lambda}\left(1 + \lambda + \dfrac{\lambda^2}{2!} + \dots\right)$, taken from left to right. Hence:

probability of no breakdowns is $e^{-\lambda} = e^{-2.1} = 0.1225$
probability of 1 breakdown is $\lambda e^{-\lambda} = 2.1 e^{-2.1} = 0.2572$
probability of 2 breakdowns is $\dfrac{\lambda^2 e^{-\lambda}}{2!} = \dfrac{2.1^2 e^{-2.1}}{2 \times 1} = 0.2700$

Thus, the probability of 1 breakdown per week is **0.2572**
The probability of less than 3 breakdowns per week is the sum of the probabilities of 0, 1 and 2 breakdowns per week,

i.e., 0.1225 + 0.2572 + 0.2700, i.e., **0.6497**

Normal distribution

37   When data is obtained, it can frequently be considered to be
a sample, (i.e. a few members), drawn at random from a large
population, (i.e. a set having many members). If the sample
number is large, it is theoretically possible to choose class intervals
which are very small, but which still have a number of members
falling within each class. A frequency polygon of this data then has
a large number of small line segments and approximates to a
continuous curve. Such a curve is called a **frequency or a
distribution curve**.

38   An extremely important symmetrical distribution curve is
called the **normal curve** and is as shown in *Figure 16.11*. This
curve can be described by a mathematical equation and is the
basis of much of the work done in more advanced statistics. Many
natural occurrences such as the heights or weights of a group of
people, the sizes of components produced by a particular machine
and the life length of certain components, approximate to a normal
distribution.

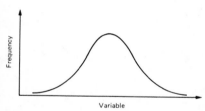

Figure 16.11

39   Normal distribution curves can differ from one another in the
following four ways:

   (a) by having different mean values,
   (b) by having different values of standard deviations,
   (c) the variables having different values and different units,
       and
   (d) by having different areas between the curve and the
       horizontal axis.

40   A normal distribution curve is **standardised** as follows:

   (a) The mean value of the unstandardised curve is made the
       origin, thus making the mean value, $\bar{x}$, zero.

(b) The horizontal axis is scaled in standard deviations. This is done by letting $z = \dfrac{x - \bar{x}}{\sigma}$, where z is called the **normal standard variate**, x is the value of the variable, $\bar{x}$ is the mean value of the distribution and σ is the standard deviation of the distribution.

(c) The area between the normal curve and the horizontal axis is made equal to unity.

When a normal distribution curve has been standardised, the normal curve is called a **standardised normal curve** or a **normal probability curve**, and any normally distributed data may be represented by the **same** normal probability curve.

41   The area under part of a normal probability curve is directly proportional to probability and the value of the shaded area shown in *Figure 16.12* can be determined by evaluating:

$$\int_{z_1}^{z_2} \frac{1}{\sqrt{(2\pi)}} e^{\left[-\frac{z^2}{2}\right]}$$

where $z = \dfrac{x - \bar{x}}{\sigma}$ (see para. 40)

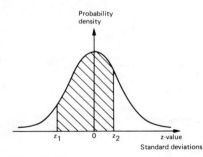

Figure 16.12

To save repeatedly determining the values of this function, tables of partial areas under the standardised normal curve are available and such a table is shown in *Table 16.3*.

For example, let the mean height of 500 people be 170 cm, and the standard deviation be 9 cm. Assuming the heights are

276

Table 16.3

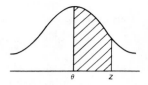

$z = \dfrac{x-\bar{x}}{\sigma}$	0	1	2	3	4	5	6	7	8	9
0.0	0.0000	0.0040	0.0080	0.0120	0.0159	0.0199	0.0239	0.0279	0.0319	0.0359
0.1	0.0398	0.0438	0.0478	0.0517	0.0557	0.0596	0.0636	0.0678	0.0714	0.0753
0.2	0.0793	0.0832	0.0871	0.0910	0.0948	0.0987	0.1026	0.1064	0.1103	0.1141
0.3	0.1179	0.1217	0.1255	0.1293	0.1331	0.1388	0.1406	0.1443	0.1480	0.1517
0.4	0.1554	0.1591	0.1628	0.1664	0.1700	0.1736	0.1772	0.1808	0.1844	0.1879
0.5	0.1915	0.1950	0.1985	0.2019	0.2054	0.2086	0.2123	0.2157	0.2190	0.2224
0.6	0.2257	0.2291	0.2324	0.2357	0.2389	0.2422	0.2454	0.2486	0.2517	0.2549
0.7	0.2580	0.2611	0.2642	0.2673	0.2704	0.2734	0.2760	0.2794	0.2823	0.2852
0.8	0.2881	0.2910	0.2939	0.2967	0.2995	0.3023	0.3051	0.3078	0.3106	0.3133
0.9	0.3159	0.3186	0.3212	0.3238	0.3264	0.3289	0.3215	0.3340	0.3365	0.3389
1.0	0.3413	0.3438	0.3451	0.3485	0.3508	0.3531	0.3554	0.3577	0.3599	0.3621
1.1	0.3643	0.3665	0.3686	0.3708	0.3729	0.3749	0.3770	0.3790	0.3810	0.3830
1.2	0.3849	0.3869	0.3888	0.3907	0.3925	0.3944	0.3962	0.3980	0.3997	0.4015
1.3	0.4032	0.4049	0.4066	0.4082	0.4099	0.4115	0.4131	0.4147	0.4162	0.4177
1.4	0.4192	0.4207	0.4222	0.4236	0.4251	0.4265	0.4279	0.4292	0.4306	0.4319
1.5	0.4332	0.4345	0.4357	0.4370	0.4382	0.4394	0.4406	0.4418	0.4430	0.4441
1.6	0.4452	0.4463	0.4474	0.4484	0.4495	0.4505	0.4515	0.4525	0.4535	0.4545
1.7	0.4554	0.4564	0.4573	0.4582	0.4591	0.4509	0.4608	0.4616	0.4625	0.4633
1.8	0.4641	0.4649	0.4656	0.4664	0.4671	0.4678	0.4686	0.4693	0.4699	0.4706
1.9	0.4713	0.4719	0.4726	0.4732	0.4738	0.4744	0.4750	0.4756	0.4762	0.4767
2.0	0.4772	0.4778	0.4783	0.4785	0.4793	0.4798	0.4803	0.4808	0.4812	0.4817
2.1	0.4821	0.4826	0.4830	0.4834	0.4838	0.4842	0.4846	0.4850	0.4854	0.4857
2.2	0.4861	0.4864	0.4868	0.4871	0.4875	0.4878	0.4881	0.4884	0.4882	0.4890
2.3	0.4893	0.4896	0.4898	0.4901	0.4904	0.4906	0.4909	0.4911	0.4913	0.4916
2.5	0.4938	0.4940	0.4941	0.4943	0.4945	0.4946	0.4948	0.4949	0.4951	0.4952
2.6	0.4953	0.4955	0.4956	0.4957	0.4959	0.4960	0.4961	0.4962	0.4963	0.4964
2.7	0.4965	0.4966	0.4967	0.4968	0.4969	0.4970	0.4971	0.4972	0.4973	0.4974
2.8	0.4974	0.4975	0.4076	0.4977	0.4977	0.4978	0.4979	0.4980	0.4980	0.4981
2.9	0.4981	0.4982	0.4982	0.4983	0.4984	0.4984	0.4985	0.4985	0.4986	0.4986
3.0	0.4987	0.4987	0.4987	0.4988	0.4988	0.4989	0.4989	0.4989	0.4990	0.4990
3.1	0.4990	0.4991	0.4991	0.4991	0.4992	0.4992	0.4992	0.4992	0.4993	0.4993
3.2	0.4993	0.4993	0.4994	0.4994	0.4994	0.4994	0.4994	0.4995	0.4995	0.4995
3.3	0.4995	0.4995	0.4995	0.4996	0.4996	0.4996	0.4996	0.4996	0.4996	0.4997
3.4	0.4997	0.4997	0.4997	0.4997	0.4997	0.4997	0.4997	0.4997	0.4997	0.4998
3.5	0.4998	0.4998	0.4998	0.4998	0.4998	0.4998	0.4998	0.4998	0.4998	0.4998
3.6	0.4998	0.4998	0.4999	0.4999	0.4999	0.4999	0.4999	0.4999	0.4999	0.4999
3.7	0.4999	0.4999	0.4999	0.4999	0.4999	0.4999	0.4999	0.4999	0.4999	0.4999
3.8	0.4999	0.4999	0.4999	0.4999	0.4999	0.4999	0.4999	0.4999	0.4999	0.4999
3.9	0.5000	0.5000	0.5000	0.5000	0.5000	0.5000	0.5000	0.5000	0.5000	0.5000

normally distributed, the number of people likely to have heights between, say, 150 cm and 195 cm is determined as follows:

The mean value $\bar{x}$ is 170 cm and corresponds to a normal standard variate value z of zero on the standardised normal curve.

A height of 150 cm has a z-value given by $z = \dfrac{x - \bar{x}}{\sigma}$ standard deviations, i.e., $\dfrac{150 - 170}{9}$ or -2.22 standard deviations. Using a table of partial areas beneath the standardised normal curve (see *Table 16.3*), a z-value of -2.22 corresponds to an area of 0.4868 between the mean value and the ordinate $z = -2.22$. The negative z-value shows that it lies to the left of the $z = 0$ ordinate. This area is shown shaded in *Figure 16.13(a)*. Similarly, 195 cm has a z-value of $\dfrac{195 - 170}{9}$ that is 2.78 standard deviations. From *Table 16.3*, this value of z corresponds to an area of 0.4973, the positive value of z showing that it lies to the right of the $z = 0$ ordinate. This area is shown shaded in *Figure 16.13(b)*. The total area shaded in *Figures 16.13(a)* and *16.13(b)* is shown in *Figure 16.13(c)* and is 0.4868 + 0.4973, i.e., 0.9841 of the total area beneath the curve.

However, from para 41, the area is directly proportional to probability. Thus, the probability that a person will have a height of between 150 and 195 cm is 0.9841. For a group of 500 people, 500×0.9841, i.e. **492 people** are likely to have heights in this range. The value of 500×0.9841 is 492.05, but since answers based on a normal probability distribution can only be approximate, results are usually given correct to the nearest whole number. Similarly, the number of people likely to have heights of less than 165 cm is determined as follows:

A height of 165 cm corresponds to $\dfrac{165 - 170}{9}$, i.e., -0.56 standard deviations. The area between $z = 0$ and $z = -0.56$ (from *Table 16.3*) is 0.2123, shown shaded in *Figure 16.14(a)*. The total area under the standardised normal curve is unity and since the curve is symmetrical, it follows that the total area to the left of the $z = 0$ ordinate is 0.5000. Thus the area to the left of the $z = -0.56$ ordinate, ('left' means 'less than', 'right' means 'more than'), is $0.5000 - 0.2123$, i.e., 0.2877 of the total area, which is shown shaded in *Figure 16.14(b)*.

From para 41, the area is directly proportional to probability and since the total area beneath the standardised normal curve is unity, the probability of a person's height being less than 165 cm is 0.2877. For a group of 500 people, 500×0.2877, i.e. **144** people are likely to have heights of less than 165 cm.

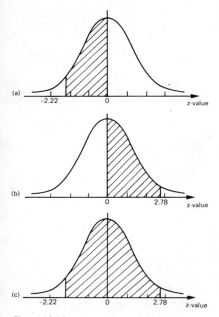

(a) -2.22 0 z-value

(b) 0 2.78 z-value

(c) -2.22 0 2.78 z-value

Figure 16.13

The number of people likely to have heights of more than 194 cm, is determined as follows:

194 cm corresponds to a z-value of $\dfrac{194-170}{9}$, that is, 2.67 standard deviations.

From *Table 16.3*, the area between $z=0$, $z=2.67$ and the standardised normal curve is 0.4962, shown shaded in *Figure 16.15(a)*. Since the standardised normal curve is symmetrical, the total area to the right of the $z=0$ ordinate is 0.5000, hence the shaded area shown in *Figure 16.15(b)* is $0.5000-0.4962$, i.e. 0.0038. From para 41, this area represents the probability of a person having a height of more than 194 cm, and for 500 people, the number of people likely to have a height of more than 194 cm is 0.0038×500, i.e. **2 people**.

279

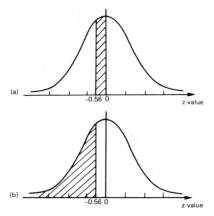

Figure 16.14

42   It should never be assumed that because data is continuous it automatically follows that it is normally distributed. One way of checking that data is normally distributed is by using **normal probability paper**, often just called **probability paper**. This is special graph paper which has linear markings on one axis and percentage probability values from 0.01 to 99.99 on the other axis. The divisions on the probability axis are such that a straight line graph results for normally distributed data when percentage cumulative frequency values are plotted against upper class boundary values.

    If the points do not lie in a reasonably straight line, then the data is not normally distributed.

43   The mean value and standard deviation of normally distributed data may be determined using normal probability paper.

    For normally distributed data, the area beneath the standardised normal curve and a z-value of unity (i.e. one standard deviation) may be obtained from *Table 16.3*. For one standard deviation, this area is 0.3413, i.e., 34.13%. An area of ± 1 standard deviation is symmetrically placed on either side of the $z = 0$ value, i.e. is symmetrically placed on either side of the 50 per cent cumulative frequency value. Thus an area corresponding to ± 1 standard deviation extends from percentage cumulative frequency values of $(50 + 34.13)\%$ to $(50 - 34.13)\%$, i.e., from 84.13% to 15.87%.

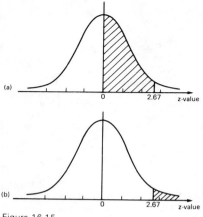

Figure 16.15

For most purposes, these values are taken as 16% and 84%. Thus, when using normal probability paper, the standard deviation of the distribution is given by:

(variable value for 84% cumulative frequency) −
  (variable value for 16% cumulative frequency)

$$\overline{}\atop{2}$$

For example, the data given below refers to the masses of 50 copper ingots. Normal probability paper may be used to determine whether the data is approximately normally distributed.

Class mid-point value (kg)	29.5	30.5	31.5	32.5	33.5	34.5	35.5	36.5	37.5	38.5
Frequency	2	4	6	8	9	8	6	4	2	1

To test the normality of a distribution, the upper class boundary/percentage cumulative frequency values are plotted on normal probability paper. The upper class boundary values are: 30, 31, 32, ..., 38, 39. The corresponding cumulative frequency values, (for 'less than' the upper class boundary values), are: 2, (4 + 2) = 6, (6 + 4 + 2) = 12, 20, 29, 37, 43, 47, 49 and 50. The corresponding percentage cumulative frequency values are

$\dfrac{2}{50} \times 100 = 4$, $\dfrac{6}{50} \times 100 = 12$, 24, 40, 58, 74, 86, 94, 98 and 100%.

$= 4$, $\dfrac{}{50} \times 100 = 12$, 24, 40, 58, 74, 86, 94, 98 and 100%.

The co-ordinates of upper class boundary/percentage cumulative frequency values are plotted as shown in *Figure 16.16*. When plotting these values, it will always be found that the co-ordinate for the 100% cumulative frequency value cannot be plotted, since the maximum value on the probability scale is 99.99. Since the points plotted in *Figure 16.16* lie very nearly in a straight line, the data is approximately normally distributed. The mean value and standard deviation can be determined from *Figure 16.16*.

Since a normal curve is symmetrical, the mean value is the value of the variable corresponding to a 50% cumulative frequency value, shown as point P on the graph. This shows that the **mean value is 33.6 kg**. The standard deviation is determined using the 84% and 16% cumulative frequency values, shown as Q and R in *Figure 16.16*. The variable values for Q and R are 35.7 and 31.4 respectively; thus two standard deviations correspond to $35.7 - 31.4$, i.e. 4.3, showing that the standard deviation of the distribution is approximately $\dfrac{4.3}{2}$, i.e. **2.15 standard deviations**.

Linear correlation

44   Correlation is a measure of the amount of association existing between two variables. For linear correlation, if points are plotted on a graph and all the points lie on a straight line, then **perfect linear correlation** is said to exist. When a straight line having a positive gradient can reasonably be drawn through points on a graph **positive or direct linear correlation** exists. Similarly, when a straight line having a negative gradient can reasonably be drawn through points on a graph, **negative or inverse linear correlation** exists. When there is no apparent relationship between co-ordinate values plotted on a graph then no correlation exists between the points.

45   The amount of linear correlation between two variables is expressed by a **coefficient of correlation**, given the symbol r. This is defined in terms of the deviations of the co-ordinates of two variables from their mean values and is given by the **product-moment formula** which states:

coefficient of correlation, $r = \dfrac{\sum xy}{\sqrt{\{(\sum x^2)(\sum y^2)\}}}$,   (1)

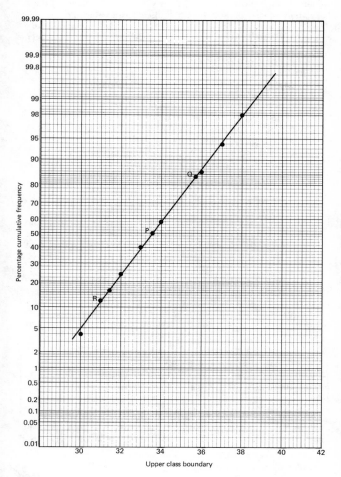

Figure 16.16

where the x-values are the values of the deviations of co-ordinates X and $\bar{X}$, their mean value and the y-values are the values of the deviations of co-ordinates Y from $\bar{Y}$, their mean value. That is, $x = (X - \bar{X})$ and $y = (Y - \bar{Y})$. The results of this determination give values of r lying between $+1$ and -1, where $+1$ indicates perfect direct correlation, -1 indicates perfect inverse correlation and 0 indicates that no correlation exists. Between these values, the smaller the value of r, the less is the amount of correlation which exists. Generally, values of r in the ranges 0.7 to 1 and -0.7 to -1 show that a fair amount of correlation exists.

46   When the value of the coefficient of correlation has been obtained from the product-moment formula, some care is needed before coming to conclusions based on this result. Checks should be made to ascertain the following two points:

(a) that a 'cause and effect' relationship exists between the variables; it is relatively easy, mathematically, to show that some correlation exists between, say, the number of ice creams sold in a given period of time and the number of chimneys swept in the same periods of time, although there is no relationship between these variables;

(b) that a linear relationship exists between the variables; the product-moment formula given in para 45 is based on linear correlation. Perfect non-linear correlation may exist, (for example, the co-ordinates exactly following the curve $y = x^3$), but this gives a low value of coefficient of correlation since the value of r is determined using the product-moment formula, based on a linear relationship.

For example, in an experiment to determine the relationship between force on a wire and the resulting extension, the following data is obtained:

Force (N)	10	20	30	40	50	60	70
Extension (mm)	0.22	0.40	0.61	0.85	1.20	1.45	1.70

The linear coefficient of correlation for this data is obtained as follows. Let X be the variable force values and Y be the dependent variable extension values. The coefficient of correlation is given by:

$$r = \frac{\sum xy}{\sqrt{\{(\sum x^2)(\sum y^2)\}}}$$

where $x = (X - \bar{X})$ and $y = (Y - \bar{Y})$, $\bar{X}$ and $\bar{Y}$ being the mean values of the X and Y values respectively. Using a tabular method to determine the quantities of this formula gives:

X	Y	$x = (X - \bar{X})$	$y = (Y - \bar{Y})$	xy	x^2	y^2
10	0.22	−30	−0.699	20.97	900	0.489
20	0.40	−20	−0.519	10.38	400	0.269
30	0.61	−10	−0.309	3.09	100	0.095
40	0.85	0	−0.069	0	0	0.005
50	1.20	10	0.281	2.81	100	0.079
60	1.45	20	0.531	10.62	400	0.282
70	1.70	30	0.781	23.43	900	0.610

$\sum X = 280$　　$\sum Y = 6.43$

$\bar{X} = \dfrac{280}{7}$　　$Y = \dfrac{6.43}{7}$　　　　$\sum xy = \quad \sum x^2 = \quad \sum y^2 =$

$= 40$　　　　$= 0.919$　　　　　71.30　　2800　　1.829

Thus $r = \dfrac{71.3}{\sqrt{[2800 \times 1.829]}} = 0.996$

This shows that a very good direct correlation exists between the values of force and extension.

Linear regression

47   Regression analysis, usually termed **regression**, is used to draw the line of 'best fit' through co-ordinates on a graph. The techniques used enable a mathematical equation of the straight line form $y = mx + c$ to be deduced for a given set of co-ordinate values, the line being such that the sum of the deviations of the co-ordinate values from the line is a minimum, i.e. it is the line of 'best fit'.

48   When a regression analysis is made, it is possible to obtain two lines of best fit, depending on which variable is selected as the dependent variable and which variable is the independent variable. For example, in a resistive electrical circuit, the current flowing is directly proportional to the voltage applied to the circuit. There are two ways of obtaining experimental values relating the current and voltage. Either, certain voltages are applied to the circuit and the current values are measured, in which case, the voltage is the independent variable and the current is the dependent variable. Alternatively, the voltage can be adjusted until a desired value of

current is flowing and the value of voltage is measured, in which case, the current is the independent value and the voltage is the dependent value.

The least-squares regression line

49   For a given set of co-ordinate values, (X_1, Y_1), (X_2, Y_2), ..., (X_N, Y_N) let the X-values be the independent variables and the Y-values be the dependent values. Also, let $D_1, D_2, ..., D_N$ be the vertical distances between the line shown as **PQ** in *Figure 16.17* and the points representing the co-ordinate values. The least-squares regression line, i.e. the line of best fit, is the line which makes the value of $D_1^2 + D_2^2 + ... D_N^2$ a minimum value.

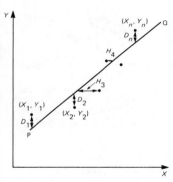

Figure 16.17

50   The equation of the least-squares regression line is usually written as $Y = a_0 + a_1X$, where a_0 is the Y-axis intercept value and a_1 is the gradient of the line (analagous to c and m in the equation $y = mx + c$). The values of a_0 and a_1 to make the sum of the 'deviations squared' a minimum can be obtained from the two equations:

$$\sum Y = a_0 N + a_1 \sum X \qquad (1)$$
$$\sum (XY) = a_0 \sum X + a_1 \sum X^2 \qquad (2)$$

where X and Y are the co-ordinate values, N is the number of co-ordinates and a_0 and a_1 are called the **regression coefficients** of Y on X. Equations (1) and (2) are called the **normal equations** of the regression line of Y on X. The regression line of Y on X is used to estimate values of Y for given values of X.

51   If the Y-values (vertical-axis), are selected as the independent variables, the horizontal distances between the line shown as PQ in *Figure 16.17* and the co-ordinate values, (H_3, H_4, etc.) are taken as the deviations. The equation of the regression line is of the form: $X = b_0 + b_1 Y$ and the normal equations become:

$$\sum X = b_0 N + b_1 \sum Y \qquad (3)$$
$$\sum (XY) = b_0 \sum Y + b_1 \sum Y^2 \qquad (4)$$

where X and Y are the co-ordinate values, b_0 and b_1 are the regression coefficients of X on Y and N is the number of co-ordinates. These normal equations are of the regression line of X on Y, which is slightly different to the regression line of Y on X. The regression line of X on Y is used to estimate values of X for given values of Y.

52   The regression line of Y on X is used to determine any value of Y corresponding to a given value of X. If the value of Y lies within the range of Y-values of the extreme co-ordinates, the process of finding the corresponding value of X is called **linear interpolation**. If it lies outside of the range of Y-values of the extreme co-ordinates then the process is called **linear extrapolation** and the assumption must be made that the line of best fit extends outside of the range of the co-ordinate values given. By using the regression line of X on Y, values of X corresponding to given values of Y may be found by either interpolation or extrapolation. For example, the experimental values relating centripetal force and radius, for a mass travelling at constant velocity in a circle, are as shown:

Force (N)	5	10	15	20	25	30	35	40
Radius (cm)	55	30	16	12	11	9	7	5

The equation of the regression line of force on radius is determined as follows:

Let the radius be the independent variable X, and the force be the dependent variable Y. (This decision is usually based on a 'cause' corresponding to X and an 'effect' corresponding to Y.)

The equation of the regression line of force on radius is of the form $Y = a_0 + a_1 X$ and the constants a_0 and a_1 are determined from the normal equations:

$$\sum Y = a_0 N + a_1 \sum X$$
$$\sum XY = a_0 \sum X + a_1 \sum X^2 \text{ (see para 50)}$$

Using a tabular approach to determine the values of the summations, gives:

Radius, X	Force, Y	X^2	XY	Y^2
55	5	3025	275	25
30	10	900	300	100
16	15	256	240	225
12	20	144	240	400
11	25	121	275	625
9	30	81	270	900
7	35	49	245	1225
5	40	25	200	1600
$\sum X = 145$	$\sum Y = 180$	$\sum X^2 = 4601$	$\sum XY = 2045$	$\sum Y^2 = 5100$

Thus $180 = 8a_0 + 145a$
$2045 = 145a_0 + 4601a_1$

Solving these simultaneous equations gives $a_0 = 33.7$ and $a_1 = -0.617$, correct to 3 significant figures. Thus the equation of the regression line of force on radius is:

$$Y = 33.7 - 0.617X$$

The equation of the regression line of radius on force is of the form $X = b_0 + b_1 Y$, and the constants b_0 and b_1 are determined from the normal equations:

$\sum X = b_0 N + b_1 \sum Y$
$\sum XY = b_0 \sum Y + b_1 \sum Y^2$, (see para 51).

The values of the summations have been obtained in the above table giving:

$145 = 8b_0 + 180b_1$
$2045 = 180b_0 + 5100b_1$

Solving these simultaneous equations gives $b_0 = 44.2$ and $b_1 = -1.16$, correct to 3 significant figures. Thus the equation of the regression line of radius on force is: $X = 44.2 - 1.16Y$.

17 Laplace transforms

1   The solution of most electrical circuit problems can be reduced ultimately to the solution of differential equations. The use of Laplace transforms provides an alternative method for solving linear differential equations.

Definition

2   The Laplace transform of the function $f(t)$ is defined by the integral

$$\int_{0}^{\infty} e^{-st} f(t) \ dt,$$

where s is a parameter assumed to be a real number.

Common notations used for the Laplace transform

3   There are various commonly used notations for the Laplace transform of $f(t)$ and these include:

   (i) $\mathcal{L}\{f(t)\}$ or $\mathscr{L}\{f(t)\}$
   (ii) $\mathcal{L}(f)$ or $\mathscr{L}f$
   (iii) $\bar{f}(s)$ or $\tilde{f}(s)$

Also, the letter p is sometimes used instead of s as the parameter. The notation adopted in this book will be $f(t)$ for the original function and $\mathcal{L}\{f(t)\}$ for its Laplace transform.
Hence, from para 2

$$\mathcal{L}\{f(t)\} = \int_{0}^{\infty} e^{-st} f(t) \ dt \qquad (1)$$

Linearity property of the Laplace transform

4   From equation (1), $\mathcal{L}\{k\,f(t)\} = \int_{0}^{\infty} e^{-st}k\,f(t)\,dt = k\int_{0}^{\infty} e^{-st}f(t)\,dt$

i.e. $\mathcal{L}\{k\,f(t)\} = k\mathcal{L}\{f(t)\}$ $\qquad\qquad$ (2)

where k is any constant.

Similarly, $\mathcal{L}\{a\,f(t) + b\,g(t)\} = \int_{0}^{\infty} e^{-st}\{a + f(t) + bg(t)\}dt$

$$= a\int_{0}^{\infty} e^{-st}f(t)\,dt + b\int_{0}^{\infty} e^{-st}g(t)\,dt$$

i.e. $\mathcal{L}\{a\,f(t) + b\,g(t)\} = a\mathcal{L}\{f(t)\} + b\mathcal{L}\{g(t)\}$,

$\qquad\qquad$ (3)

where a and e any real constants.

The Laplace transform is termed a **linear operator** because of the properties shown in equations (2) and (3). Thus, for example,

$$\mathcal{L}\left\{1 + 2t - \frac{1}{3}t^4\right\} = \mathcal{L}\{1\} + 2\mathcal{L}\{t\} - \frac{1}{3}\mathcal{L}\{t^4\}$$

from equations (2) and (3) of para 4

$$= \frac{1}{s} + 2\left(\frac{1}{s^2}\right) - \frac{1}{3}\left(\frac{4!}{s^{4+1}}\right)$$

from (i), (vi) and (viii) of *Table 17.1*.

$$= \frac{1}{s} + \frac{2}{s^2} - \frac{8}{s^5}$$

$$\mathcal{L}\{5e^{2t} - 3e^{-t}\} = 5\mathcal{L}\{e^{2t}\} - 2\mathcal{L}\{e^{-t}\}$$

$$= 5\left(\frac{1}{s-2}\right) - 3\left(\frac{1}{s--1}\right)$$

from (iii) of *Table 17.1*.

$$= \frac{5}{s-2} - \frac{3}{s+1} = \frac{5(s+1) - 3(s-2)}{(s-2)(s+1)}$$

$$= \frac{2s+11}{(s-2)(s+1)}$$

Table 17.1 *Elementary standard Laplace transforms*

Function $f(t)$	Laplace transforms $\{f(t)\} = \int_0^\infty e^{-st} f(t)\, dt$
(i) $\quad 1$	$\dfrac{1}{s}$
(ii) $\quad k$	$\dfrac{k}{s}$
(iii) $\quad e^{at}$	$\dfrac{1}{s-a}$
(iv) $\quad \sin at$	$\dfrac{a}{s^2+a^2}$
(v) $\quad \cos at$	$\dfrac{s}{s^2+a^2}$
(vi) $\quad t$	$\dfrac{1}{s^2}$
(vii) $\quad t^2$	$\dfrac{2!}{s^3}$
(viii) $\quad t^n \,(n=1,2,3,\ldots)$	$\dfrac{n!}{s^{n+1}}$
(ix) $\quad \cosh at$	$\dfrac{s}{s^2-a^2}$
(x) $\quad \sinh at$	$\dfrac{a}{s^2-a^2}$
(xi) $\quad e^{at}\, t^n$	$\dfrac{n!}{(s-a)^{n+1}}$
(xii) $\quad e^{at} \sin \omega t$	$\dfrac{\omega}{(s-a)^2+\omega^2}$
(xiii) $\quad e^{at} \cos \omega t$	$\dfrac{s-a}{(s-a)^2+\omega^2}$
(xiv) $\quad e^{at} \sinh \omega t$	$\dfrac{\omega}{(s-a)^2-\omega^2}$
(xv) $\quad e^{at} \cosh \omega t$	$\dfrac{s-a}{(s-a)^2-\omega^2}$

$$\mathscr{L}\{6 \sin 3t - 4 \cos 5t\} = 6\mathscr{L}\{\sin 3t\} - 4\mathscr{L}\{\cos 5t\}$$

$$= 6\left(\frac{3}{s^2 + 3^2}\right) - 4\left(\frac{s}{s^2 + 5^2}\right)$$

from (iv) and (v) of *Table 17.1*.

$$= \frac{18}{s^2 + 9} - \frac{4s}{s^2 + 25}$$

$$\mathscr{L}\{2t^4 e^{3t}\} = 2\mathscr{L}\{t^4 e^{3t}\} = 2\left(\frac{4!}{(s-3)^{4+1}}\right) = \frac{48}{(s-3)^5}$$

from (xi) of *Table 17.1*.

$$\mathscr{L}\{4e^{3t} \cos 5t\} = 4\mathscr{L}\{e^{3t} \cos 5t\} = 4\left(\frac{s-3}{(s-3)^2 + 5^2}\right)$$

from (xiii) of *Table 17.1*.

$$= \frac{4(s-3)}{s^2 - 6s + 9 + 25} = \frac{4(s-3)}{s^2 - 6s + 34}$$

$$\mathscr{L}\{5e^{-3t} \sinh 2t\} = 5\left(\frac{2}{(s--3)^2 - 2^2}\right)$$

from (xiv) of *Table 17.1*.

$$= \frac{10}{(s+3)^2 - 2^2} = \frac{10}{s^2 + 6s + 9 - 4} = \frac{10}{s^2 + 6s + 5}$$

Laplace transforms of derivatives

(a) *First derivative*

6   Let the first derivative of $f(t)$ be $f'(t)$ then, from equation (1),

$$\mathscr{L}\{f'(t)\} = \int_0^\infty e^{-st} f'(t) \, dt$$

From page 194, when integrating by parts

$$\int u \frac{dv}{dt} dt = uv - \int v \frac{du}{dt} dt$$

When evaluating

$$\int_0^\infty e^{-st} f'(t) \, dt, \text{ let } u = e^{-st} \text{ and } \frac{dv}{dt} = f'(t),$$

from which,

$$\frac{du}{dt} = -se^{-st} \text{ and } v = \int f'(t) = f(t)$$

Hence

$$\int_0^\infty e^{-st} f'(t) \ dt = (e^{-st} f(t))_0^\infty + \int_0^\infty f(t)(-se^{-st}) \ dt$$

$$= [0 - f(0)] + s \int_0^\infty e^{-st} f(t) \ dt$$

$$= -f(0) + s\mathcal{L}\{f(t)\}, \text{ assuming } e^{-st}f(t) \rightarrow 0$$

as $t \rightarrow 0$, and $f(0)$ is the value of $f(t)$ at $t = 0$.

Hence $\mathcal{L}\{f'(t)\} = s\mathcal{L}\{f(t)\} - f(0)$ (4)

or $\mathcal{L}\left\{\dfrac{dy}{dx}\right\} = s\mathcal{L}\{y\} - y(0)$

where $y(0)$ is the value of y at $x = 0$.

(b) Second derivative

Let the second derivative of $f(t)$ be $f''(t)$, then from equation (1),

$$\mathcal{L}\{f''(t)\} = \int_0^\infty e^{-st} f''(t) \ dt$$

Integrating by parts gives:

$$\int_0^\infty e^{-st} f''(t) \ dt = [e^{-st} f']_0^\infty + s \int_0^\infty e^{-st} f'(t) \ dt$$

$$= [0 - f'(0)] + s\mathcal{L}\{f'(t)\}$$

assuming $e^{-st}f'(t) \rightarrow 0$ as $t \rightarrow 0$, and $f'(0)$ is the value of $f'(t)$ at $t = 0$. Hence

$$\angle\{f'(t)\} = -f'(0) + s[s\mathcal{L}\{f(t)\} - f(0)],$$

from equation (4),

i.e., $\mathcal{L}\{f'(t)\} = s^2\mathcal{L}\{f(t)\} - sf(0) - f'(0)$

or $\mathscr{L}\left\{\dfrac{d^2y}{dx^2}\right\} = s^2 \mathscr{L}\{y\} - sy(0) = y'(0)$

where $y'(0)$ is the value of $\dfrac{dy}{dx}$ at $x = 0$.

Equations (4) and (5) are important and are used in the solution of differential equations (see para 10).

Initial value and final value theorem

7   There are several Laplace transform theorems used to simplify and interpret the solution of certain problems. Two such theorems are the initial value theorem and the final value theorem.

(a) The **initial value theorem** states:

$$\underset{t \to 0}{\text{limit}}\,[f(t)] = \underset{s \to \infty}{\text{limit}}\,[s\,\mathscr{L}\{f(t)\}]$$

(b) The **final value theorem** states:

$$\underset{t \to \infty}{\text{limit}}\,[f(t)] = \underset{s \to 0}{\text{limit}}\,[s\,\mathscr{L}\{f(t)\}]$$

For example, if $f(t) = 3e^{4t}$ then:

$$\underset{t \to \infty}{\text{limit}}\,[3e^{4t}] = \underset{s \to 0}{\text{limit}}\left[s\left(\dfrac{3}{s-4}\right)\right]$$

i.e., $3e^{\infty} = 0\left(\dfrac{3}{0-4}\right)$

i.e., $\mathbf{0 = 0}$, which illustrates the theorem.

The initial and final value theorems are used in pulse circuit applications where the response of the circuit for small periods of time, or the behaviour immediately after the switch is closed are of interest. The final value theorem is particularly useful in investigating the stability of systems (such as in automatic aircraft-landing systems) and is concerned with the steady state response for large values of time t, i.e. after all transient effects have died away.

Inverse Laplace transforms

8   If the Laplace transform of a function $f(t)$ is $F(s)$, i.e.

$$\mathscr{L}\{f(t)\} = F(s)$$

then $f(t)$ is called the **inverse Laplace transform** of $F(s)$ and is written as

$$f(t) = \mathscr{L}^{-1}\{F(s)\}.$$

For example, since

$$\mathscr{L}\{1\} = \frac{1}{s} \text{ then } \mathscr{L}^{-1}\left\{\frac{1}{s}\right\} = 1$$

Similarly, since

$$\mathscr{L}\{\sin at\} = \frac{a}{s^2 + a^2} \text{ then } \mathscr{L}^{-1}\left\{\frac{a}{s^2 + a^2}\right\} = \sin at,$$

and so on.

Tables of Laplace transforms (such as *Table 17.1* in para 5) may be used to find inverse Laplace transforms. Thus, for example,

$$\mathscr{L}^{-1}\left\{\frac{1}{s^2 + 9}\right\} = \mathscr{L}^{-1}\left\{\frac{1}{s^2 + 3^2}\right\} = \frac{1}{3}\sin 3t,$$

from (iv) of *Table 17.1*.

$$\mathscr{L}^{-1}\left\{\frac{5}{3s - 1}\right\} = \mathscr{L}^{-1}\left\{\frac{5}{\left(s - \frac{1}{3}\right)}\right\} = \frac{5}{3}e^{(1/3)t}$$

from (iii) of *Table 17.1*.

$$\mathscr{L}^{-1}\left\{\frac{3}{s^4}\right\} = \frac{1}{2}\mathscr{L}^{-1}\left\{\frac{3!}{s^4}\right\} = \frac{1}{2}t^3$$

from (viii) of *Table 17.1*.

$$\mathscr{L}^{-1}\left\{\frac{7s}{s^2 + 4}\right\} = 7\mathscr{L}^{-1}\left\{\frac{s}{s^2 + 2^2}\right\} = 7\cos 2t$$

from (v) of *Table 17.1*.

$$\mathscr{L}^{-1}\left\{\frac{3}{s^2 - 7}\right\} = 3\mathscr{L}^{-1}\left\{\frac{1}{s^2 - (\sqrt{7})^2}\right\} = \frac{3}{\sqrt{7}}\sinh\sqrt{7}t$$

from (x) of *Table 17.1*.

$$\mathscr{L}^{-1}\left\{\frac{3}{s^2 - 4s + 13}\right\} = \mathscr{L}^{-1}\left\{\frac{3}{(s - 2)^2 + 3^2}\right\} = e^{2t}\sin 3t$$

from (xii) of *Table 17.1*.

9   Sometimes the function whose inverse is required is not recognisable as a standard type, such as those listed in para 5. In

such cases it may be possible, by using **partial functions**, to resolve the function into simpler fractions which may be inverted on sight. For example, the function,

$$F(s) = \frac{2s-3}{s(s-3)}$$

cannot be inverted on sight from *Table 17.1* in para 5. However, by using partial fractions,

$$\frac{2s-3}{s(s-3)} = \frac{1}{s} + \frac{1}{s-3},$$

which may be inverted as $1 + e^{3t}$ from (i) and (iii) of *Table 17.1* in para 5. (For a summary of the forms of partial fractions, see page 35.)

Procedure to solve differential equations by using Laplace transforms

10   (i) Take the Laplace transform of both sides of the differential equation by applying the formulae for the Laplace transforms of derivatives (i.e. equations (4) and (5) of para 6) and, where necessary, using a list of standard Laplace transforms, such as *Table 17.1* in para 5.
   (ii) Put in the given initial conditions, i.e. $y(0)$ and $y'(0)$.
   (iii) Rearrange the equation to make $\mathscr{L}\{y\}$ the subject.
   (iv) Determine y by using, where necessary, partial fractions, and taking the inverse of each term.

Thus, for example, to solve

$$2\frac{d^2y}{dx^2} + 5\frac{dy}{dx} - 3y = 0$$

given that when $x=0$, $y=4$ and $\dfrac{dy}{dx}=9$, using the above procedure:

   (i) $2\mathscr{L}\left\{\dfrac{d^2y}{dx^2}\right\} + 5\,\mathscr{L}\left\{\dfrac{dy}{dx}\right\} - 3\,\mathscr{L}\{y\} = \mathscr{L}\{0\}$

   $2[s^2\,\mathscr{L}\{y\} - xy(0) - y'(0)] + 5[s\,\mathscr{L}\{y\} - y(0)] - 3\,\mathscr{L}\{y\} = 0,$

   from equations (4) and (5) of para 6.
   (ii) $y(0)=4$ and $y'(0)=9$.
   Thus $2[s^2\,\mathscr{L}\{y\} - 4s - 9] + 5[s\,\mathscr{L}\{y\} - 4] - 3\,\mathscr{L}\{y\} = 0$
   i.e. $2s^2\,\mathscr{L}\{y\} - 8s - 18 + 5s\,\mathscr{L}\{y\} - 20 - 3\,\mathscr{L}\{y\} = 0$
   (iii) **Rearranging gives:** $(2s^2 + 5s - 3)\,\mathscr{L}\{y\} = 8s + 38$

i.e. $\mathcal{L}\{y\} = \dfrac{8s+38}{2s^2+5s-3}$

(iv) $y = \mathcal{L}^{-1}\left\{\dfrac{8s+38}{2s^2+5s-3}\right\}$

$$\dfrac{8s+38}{2s^2+5s-3} = \dfrac{8s+38}{(2s-1)(s+3)} = \dfrac{A}{2s-1} + \dfrac{B}{s+3} = \dfrac{A(s+3)\,3B(2s-1)}{(2s-1)(s+3)}$$

Hence $8s+38 = A(s+3) + B(2s-1)$

When $s = \dfrac{1}{2}$, $42 = 3\dfrac{1}{2}A$, from which, $A = 12$

When $s = -3$, $14 = -7B$, from which $B = -2$

Hence $y = \mathcal{L}^{-1}\left\{\dfrac{8s+38}{2x^2+5s-3}\right\} = \mathcal{L}^{-1}\left\{\dfrac{12}{2s-1} - \dfrac{2}{s+3}\right\}$

$$= \mathcal{L}^{-1}\left\{\dfrac{12}{2\left(s-\dfrac{1}{2}\right)}\right\} - \mathcal{L}^{-1}\left\{\dfrac{2}{s+3}\right\}$$

Hence $y = 6e^{-(1/2)x} - 2e^{-3x}$, from (iii) of *Table 17.1*.

Similarly, to solve $\dfrac{d^2y}{dx^2} - 7\dfrac{dy}{dx} + 10y = e^{2x} + 20$, given that when $x = 0$,

$y = 0$ and $\dfrac{dy}{dx} = \dfrac{1}{3}$, using the above procedure:

(i) $\mathcal{L}\left\{\dfrac{2^2y}{dx^2}\right\} - 7\mathcal{L}\left\{\dfrac{dy}{dx}\right\} + 10\mathcal{L}\{y\} = \mathcal{L}\{e^{2x}+20\}$

Hence $[s^2\mathcal{L}\{y\} - sy(0) - y'(0)] - 7[s\mathcal{L}\{y\} - y(0)] + 10\mathcal{L}\{y\}$

$$= \dfrac{1}{s-2} + \dfrac{20}{s}$$

(ii) $y(0) = 0$ and $y'(0) = -\dfrac{1}{3}$

Hence $s^2\mathcal{L}\{y\} - 0 - \left(-\dfrac{1}{3}\right) - 7s\mathcal{L}\{y\} + 0 + 10\mathcal{L}\{y\}$

$$= \dfrac{21s-40}{s(s-2)}$$

(iii) $(s^2 - 7s + 10)\mathcal{L}\{y\} = \dfrac{21s-40}{s(s-2)} - \dfrac{1}{3} = \dfrac{3(21s-40) - s(s-2)}{3s(s-2)}$

$$= \dfrac{s^2 + 65s - 120}{3s(s-2)}$$

$$\mathscr{L}\{y\} = \frac{-s^2 + 65s - 120}{3s(s-2)(s^2 - 7s + 10)} = \frac{1}{3}\left[\frac{-s^2 + 65s - 120}{s(s-2)(s-2)(s-5)}\right]$$

$$= \frac{1}{3}\left[\frac{-s^2 + 65s - 120}{s(s-5)(s-2)^2}\right]$$

(iv) $\quad y = \frac{1}{3}\mathscr{L}^{-1}\left\{\frac{-s^2 + 65s - 120}{s(s-5)(s-2)^2}\right\}$

$$\frac{-s^2 + 65s - 120}{s(s-5)(s-2)^2} \equiv \frac{A}{s} + \frac{B}{s-5} + \frac{C}{s-2} + \frac{D}{(s-2)^2}$$

$$\equiv \frac{A(s-5)(s-2)^2 + B(s)(s-2)^2 + C(s)(s-5)(s-2) + D(s)(s-5)}{s(s-5)(s-2)^2}$$

Hence $-s^2 + 65s - 120 \equiv A(s-5)(s-2)^2 + B(s)(s-2)^2 +$
$$C(s)(s-5)(s-2) + D(s)(s-5)$$

When $s = 0$, $-120 = -20A$, from which, $A = 6$
When $s = 5$, $180 = 45B$, from which, $B = 4$
When $s = 2$, $6 = -6D$, from which, $D = -1$
Equating s^3 terms gives: $0 = A + B + C$, from which, $C = -10$

Hence $\frac{1}{3}\mathscr{L}^{-1}\left\{\frac{-s^2 + 65s - 120}{s(s-5)(s-2)^2}\right\} = \frac{1}{3}\mathscr{L}^{-1}\left\{\frac{6}{s} + \frac{4}{s-5} - \frac{10}{s-2} - \frac{1}{(s-2)^2}\right\}$

$$= \frac{1}{3}[6 + 4e^{5x} - 10e^{2x} - xe^{2x}]$$

Thus $y = 2 + \frac{4}{3}e^{5x} - \frac{10}{3}e^{2x} - \frac{x}{3}e^{2x}$

Index